Wi-Fi Telephony

Wi-Fi Telephony

Challenges and Solutions for Voice over WLANs

By

Praphul Chandra and
David Lide

ELSEVIER

AMSTERDAM • BOSTON • HEIDELBERG • LONDON
NEW YORK • OXFORD • PARIS • SAN DIEGO
SAN FRANCISCO • SINGAPORE • SYDNEY • TOKYO
Newnes is an imprint of Elsevier

Newnes

Newnes is an imprint of Elsevier
30 Corporate Drive, Suite 400, Burlington, MA 01803, USA
Linacre House, Jordan Hill, Oxford OX2 8DP, UK

∞ Recognizing the importance of preserving what has been written,
Elsevier prints its books on acid-free paper whenever possible.

Library of Congress Cataloging-in-Publication Data
Chandra, Praphul.
 Wi-Fi telephony : challenges and solutions for voice over WLANs / by
Praphul Chandra and David Lide.
    p. cm.
 Includes index.
 ISBN-13: 978-0-7506-7971-8 (pbk. : alk. paper)
 ISBN-10: 0-7506-7971-9 (pbk. : alk. paper) 1. Internet telephony 2.
Wireless LANs. I. Lide, David R., 1928- II. Title. III. Title: Challenges
and solutions for voice over WLANs.
 TK5105.8865.C47 2007
 004.69--dc22

                              2006027814

British Library Cataloguing-in-Publication Data
A catalogue record for this book is available from the British Library.

For information on all Newnes publications,
visit our website at www.books.elsevier.com

Transferred to Digital Printing 2009

This book is dedicated—

To my parents
&
To my wife, Shilpy.

—Praphul Chandra

To my parents.

—David Lide

Contents

Acknowledgments

I started writing this book with Dave while I was working for Texas Instruments. Since then, I have moved on and joined HP Labs, India. The separation in distance (and in time zones) has been a challenge for both of us and for our editors. I would like to thank Dave for his commitment and initiative, and our editors for being patient with us. I would also like to thank my extended family in Saharanpur, Kanpur and Datia for their constant encouragement and support. Finally, I would like to thank my friend Ashwin, who has always encouraged me to shoot for the stars.

—Praphul Chandra

I'd like to thank my colleague, Praphul Chandra, for inviting me to join him in this project and for his leadership, despite the challenges of time and distance. I'd also like to thank my family, especially my wife Nellie, for giving me the time to work on this project. Finally, I'd like to thank all my colleagues at Texas Instruments for their dedication in striving to make *Voice over Wi-Fi* a reality.

—David Lide

Acronyms

Symbol

2G	Second Generation
3G	Third Generation

A

ACL	Asynchronous Connectionless
ACM	Address-complete Message
AES	Advanced Encryption Standard
AH	Authentication Header
AIC	Analog Interface Codec
AID	Association ID
AMPS	Advanced Mobile Phone System
ANM	Answer Message
AP	Access Point
ARP	Address Resolution Protocol
AuC	Authentication Center

B

BC	Back-off Counter
BS	Base Station
BSA	Basic Service Area
BSC	Base Station Controller
BSS	Base Station Subsystem
BSS	Basic Service Set
BT	Bluetooth
BTS	Base Transceiver Station

C

CAS	Channel Associated Signaling
CBC	Cipher Block Chaining
CCK	Complementary Code Keying

CCMP	Counter Mode CBC-MAC Protocol
CCS	Common Channel Signaling
CD	Codependent Devices
CD	Collision Detection
CDMA	Code Division Multiple Access
CEPT	Conference of European Postal and Telecommunication
CFB	Contention-free Burst
CHAP	Challenge Handshape Authentication Protocol
CMR	Codec Mode Request
CMS	Call-management Servers
CO	Central Office
CP	Contention Period
CPE	Customer Premises Equipment
CRC	Cyclic Redundancy Check
CRC-32	Cyclic Redundancy Check-32 Bits
CSMA-CA	Carrier Sense Multiple Access with Collision Avoidance
CSMA-CD	Carrier Sense Multiple Access with Collision Detection
CTS	Clear To Send

D

DA	Destination Address
DARPA	Defense Department Special Projects Agency
DCF	Distributed Coordination Function
DH	Diffie-Hellman
DHCP	Dynamic Host Configuration Protocol
DIFS	DCF Inter-Frame Spacing
DNS	Domain Name System
DoS	Denial of Service
DS	Differentiated Service
DS	Distribution System
DSAP	Destination Service Access Point
DSCP	Differentiated Service Code Point
DSSS	Direct Sequence Spread Spectrum
DTMF	Dual-tone Multifrequency

E

EAP	Extensible Authentication Protocol
EAPoL	Extensible Authentication Protocol over Lan
EDCF	Enhanced Distributed Coordination Function
EIFS	Extended IFS

EIR	Equipment Identity Register
EOSP	End of Service Period
ESP	Encapsulating Security Payload
ESS	Extended Service Set

F

FCS	Frame Check Sequence
FDMA	Frequency Division Multiple Access
FDQN	Fully Qualified Domain Name
FHSS	Frequency Hopping Spread Spectrum
FMS	Fluhrer-Mantin-Shamir
FSK	Frequency Shift Keying
FTP	File Transport Protocol

G

GMSC	Gateway Mobile Switching Center
GMSK	Gaussian Minimum Shift Keying
GPRS	General Packet Radio Service
GSM	Global Systems for Mobile Communications

H

HCF	Hybrid Coordination Function
HLR	Home Location Register
HTML	HyperText Markup Language
HTTP	HyperText Transfer Protocol

I

IAM	Initial Address Message
IAPP	Inter Access Point Protocol
IBSS	Independent Basic Service Set
ICMP	Internet Control Message Protocol
ICV	Integrity Check Value
ID	Independent Devices
IE	Information Element
IFS	Inter-Frame Spacing
IGMP	Internet Group Management Protocol
IKE	Internet Key Exchange
IMS	IP Multimedia Subsystem
IMSI	International Mobile Subscriber Identity
IP	Internet Protocol

IPP	IP PHONE
IPsec	Internet Protocol Security
IS41	Interim Standard 41
ISDN	Integrated Services Data Network
ITS	Intelligent Transportation System
IV	Initialization Vector

L

LAN	Local Area Network
LDO	Low Drop-out Oscillator
LLC	Logical Link Control
LS	Land Station
LSAP	LLC Service Access Point

M

MAC	Media Access Control
MAC	Message Authentication Code
MBWA	Mobile Broadband Wireless Access
MCU	Multipoint Control Unit
ME	Mobile Equipment
MF	Multifrequency
MGCP	Media Gateway Control Protocol
MIC	Message Integrity Check
MIMO	Multiple Input, Multiple Output
MK	Master Key
MKI	Master Key Index
MPDU	Media Access Control Protocol Data Unit
MS	Mobile Station
MSC	Mobile Switching Center
MSDU	Media Access Control Service Data Unit
MSRN	Mobile Station Roaming Number
MSS	Maximum Segment Size
MTBA	Multiple TID Block ACK
MTSO	Mobile Telephone Switching Office

N

NAT	Network Address Translation
NAV	Network Allocation Vector
NCS	Network Controlled Signaling

O

OFDM	Orthogonal-Frequency-Division-Multiplexing
OOB	Out-of-Band
OSA	Open System Authentication
OSI	Open Systems Interconnection
OUI	Organizationally Unique Identifier

P

PAP	Password Authentication Protocol
PBCC	Packet Binary Convolutional Coding
PC	Point Coordinator
PCF	Point Coordination Function
PCM	Pulse-code Modulation
PESQ	Perceptual Evaluation of Voice Quality
PF	Persistence Factor
PFC	Point Coordination Function
PHY	Physical Layer
PID	Protocol Identifier
PKI	Public Key Infrastructure
PLC	Packet Loss Concealment
PLCP	Physical Layer Convergence Protocol
PMD	Physical Medium Dependent
PMK	Pair-wise Master Key
PMM	Power-management Module
PN	Packet Number
PRF	Pseudorandom Function
PSK	Phase Shift Keying
PSMP	Power Save Multi Poll
PSTN	Public Switched Telephone Network
PTK	Pair-wise Transient Key

Q

QAM	Quadrature Amplitude Modulation
QoS	Quality of Service

R

RA	Receiver Address
RADIUS	Remote Access Dial In User Security
REL	Release Message
RF	Radio Frequency

RG	Remote Gateway
RNC	Radio Network Controller
RSA	Rivest-Shamir-Adleman
RSN	Robust Security Network
RSS	Received Signal Strength
RSSI	Received Signal Strength Indication
RTC	Real-time Clock
RTCP	Real-Time Control Protocol
RTP	Real-Time Transport Protocol
RTS	Request To Send

S

S-APSD	Scheduled Automatic Power Save Delivery
SAR	Security-aware Ad Hoc Routing
SCO	Synchronous Connection-oriented
SDP	Session Description Protocol
SFD	Start Frame Delimiter
SGW	Secure Gateway
SID	System Identifier
SIFS	Short Inter-Frame Space
SIM	Subscriber Identity Module
SIP	Session Initiation Protocol
SKA	Shared Key Authentication
SMTP	Simple Mail Transport Protocol
SNR	Signal-to-Noise Ratio
SoC	System-on-Chip
SPI	Security Parameter Index
SS	System States
SS7	Signaling System #7
SSAP	Source Service Access Point
SSID	Service Set Identifier
SSL	Secure Sockets Layer
SSP	Service Switching Point
STA	Station
STP	Signaling Transfer Point

T

TA	Transmitter Address
TBTT	Target Beacon Transmission Time
TC	Traffic Category

TCP	Transmission Control Protocol
TDMA	Time Division Multiple Access
TIM	Traffic Indication Map
TKIP	Temporal Key Integrity Protocol
TLS	Transport Layer Security
TOS	Type of Service
TPC	Transmit Power Control
TSC	TKIP Sequence Counter
TSN	Transitional Security Network
TSPEC	Traffic Specifications
TXOP	Transmission Opportunity

U

U-APSD	Unscheduled-Automatic Power Save Delivery
UDP	User Datagram Protocol
UDVM	Universal Decompressor Virtual Machine
UMA	Unlicensed Mobile Access
UMTS	Universal Mobile Telecommunications System
UPSD	Unscheduled Power Save Delivery

V

VAD	Voice Activity Detection
VF	Voice Frequency
VLAN	Virtual LAN
VLR	Visitor Location Register
VoIP	Voice over IP
VPN	Virtual Private Network

W

WAN	Wide Area Network
WAP	Wireless Application Protocol
WDS	Wireless Distribution System
WEP	Wired Equivalent Privacy
WIPP	Wireless IP Phone
WLAN	Wireless Local Area Network
WME	WLAN Multimedia Enhancement
WMM-SA	Wi-Fi MultiMedia-Scheduled Access
WPA	Wi-Fi Protected Access
WRAN	Wireless Regional Area Network

About the Authors

Praphul Chandra currently works as a Senior Research Scientist at HP Labs, India which focuses on "technological innovation for emerging countries." He is an Electrical Engineer by training, though his recent interest in social science and politics has prompted him to explore the field of Public Policy. He lives with his family in Bangalore and maintains his personal website at *www.thecofi.net*.

David Lide currently is a Senior Member of the Technical Staff at Texas Instruments and has worked on various aspects of Voice over IP for the past eight years. Prior to that, he has worked on Cable Modem design and on weather satellite ground systems. He lives with his family in Rockville, Maryland.

The Telephony World

1.1 The Basics

This is a book about using wireless local area networks (LANs) to carry human speech and voice. In this first chapter, we look at how voice has traditionally been carried over networks. We begin by understanding the basic nature of human speech, using Wikipedia definitions:

> "*Sound* is a disturbance of mechanical energy that propagates through matter as a wave. Humans perceive sound by the sense of hearing. By sound, we commonly mean the vibrations that travel through air and can be heard by humans. Sound propagates as waves of alternating pressure, causing local regions of compression and rarefaction. Particles in the medium are displaced by the wave and oscillate. As a wave, sound is characterized by the properties of waves including frequency, wavelength, period, amplitude and velocity or speed."

Figure 1.1 is a schematic representation of hearing.

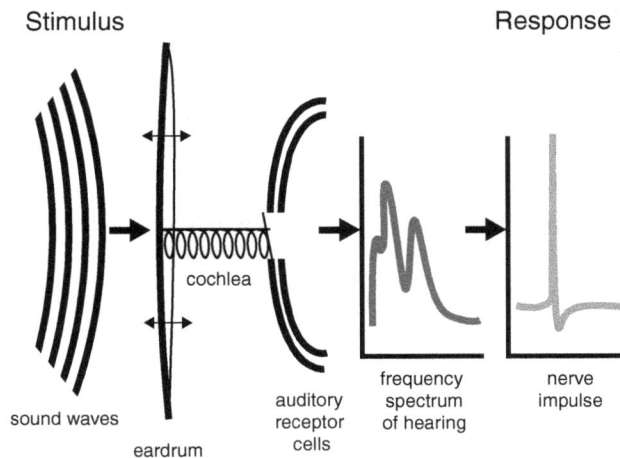

Figure 1.1: Human Hearing

"*Human voice* consists of sound made by a person using the vocal folds for talking, singing, laughing, screaming or crying. The vocal folds, in combination with the teeth, the tongue, and the lips, are capable of producing highly intricate arrays of sound, and vast differences in meaning can often be achieved through highly subtle manipulation of the sounds produced (especially in the expression of language). A voice frequency (VF) or voice band is one of the frequencies, within part of the audio range that is used for the transmission of speech. In telephony, the usable voice frequency band ranges from approximately 300 Hz to 3400 Hz. The bandwidth allocated for a single voice-frequency transmission channel is usually 4 kHz, including guard bands, allowing a sample rate of 8 kHz to be used as the basis of the pulse-code modulation system used for the digital PSTN." (PSTN is the abbreviation for Public Switched Telephone Network.)

1.1.1 The Evolution of the Telephone Network

The discovery of the telephone can be attributed to Alexander Graham Bell who in 1876 discovered that if a battery is applied across an electrical circuit (the wires) while the user speaks, the sound wave produced by the human voice could be carried across this same pair of wires to a receiving end set up to accept this electrical current and convert the electricity back into sound.

Within a few decades (NOT a long duration at that time) of Bell's discovery, the first telephone sets were being sold. The first telephone sets were sold in pairs: each telephone was connected to one and only one other telephone via a dedicated wire. This meant that if I wanted the capability to be able to call 10 people, I had to have 10 telephones on my desk. Furthermore, each telephone came with its own battery and a crank used to ring the far-end telephone. Obviously, this was not a very scaleable model.

Hence, the next step in the evolution was the development of the central office. In this model, a user needed only one telephone set, which was connected by a single wire to the central office. This reduced the demand on the infrastructure dramatically. To use the telephone, the user would simply pick up the phone handset. This would connect him to the human operator sitting at the central office. The user would then tell the human operator who he wished to be connected to and the operator would use a patch-cord system on the telephone panel to connect him to the destination party. Though much more efficient and scaleable than the one-to-one model, the model was limited in its capacity because of the human intervention required.

As the demand for telephone service grew and technology evolved, digital computers eventually replaced the manual operators. This not only increased the speed of switching but also led to an increase in the effective capacity of the network.

This eventually led to the evolution of the telephone network, aka PSTN, in its current form. For this to happen, the analog voice signal needs to be converted to the digital world.

1.2 Digitizing Speech

The human voice produces an analog signal. When a speaker pushes air out of the lungs through the glottis, air pulses escape through the mouth and sometimes the nose. These pulses produce small variations in air pressure that result in an analog signal.

Human speech can be represented as an analog wave that varies over time and has a smooth, continuous curve. The height of the wave represents intensity (loudness), and the shape of the wave represents frequency (pitch). The continuous curve of the wave accommodates an infinity of possible values. A computer must convert these values into a set of discrete values, using a process called digitization. Once speech is digitized, a computer can store speech on a hard drive and transmit speech across digital networks, including corporate networks, the Internet, and telephone-company networks, which are increasingly using digital components.

To digitize speech, an analog-digital converter samples the value of the analog signal repeatedly and encodes each result in a set of bits. In conventional PSTN telephony, before sampling, the converter filters the signal so that most of it lies between 300 and 3400 Hz. This exploits the fact that, while humans can hear frequencies as high as 20 kHz, most of the information conveyed in speech does not exceed 4 kHz.[1]

The sampling process uses a theorem developed by the American physicist Harry Nyquist in the 1920s. Nyquist's Theorem states that the sampling frequency must be at least twice as high as the highest input frequency for the result to closely resemble the original signal. Thus, the "filtered" voice signal is sampled at 8000 Hz so that frequencies up to 4000 Hz can be recorded. Every 125 µs (1/8000th of a second), the value (magnitude) of the analog voice signal is recorded as a digital value. This value is typically a number between 0 and 255 (i.e., 8 bits, which is the basic unit of storage on modern-day computers). Ten, 12 and 16 bit sampling is also popular. By sampling this often, the result is a faithful representation of the original signal, and the human ear will not hear distortion.[2]

---

[1] A hertz, or Hz, is a unit of frequency equal to one cycle per second.

[2] As a side note, in cellular and voice over IP telephony systems, 16,000-Hz sampling rate is gaining popularity. We will discuss this more in Chapter 3.

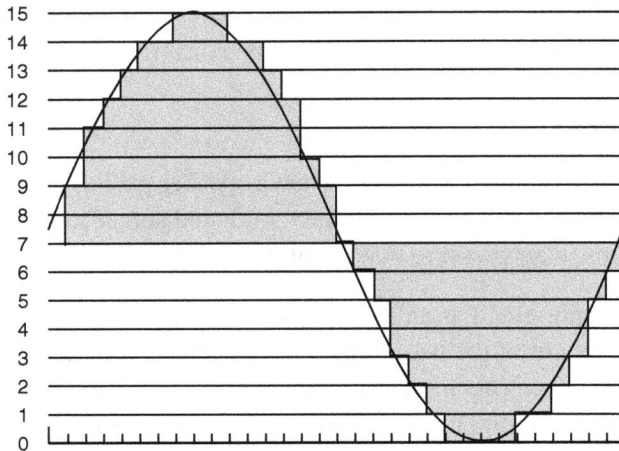

Figure 1.2: Quantization: A-D Conversion

As the digital samples are collected, modern telephony systems may convert them into a digital representation using pulse-code modulation or PCM. From Wikipedia, "Pulse-code modulation (PCM) is a digital representation of an analog signal where the magnitude of the signal is sampled regularly at uniform intervals, then quantized to a series of symbols in a digital (usually binary) code."

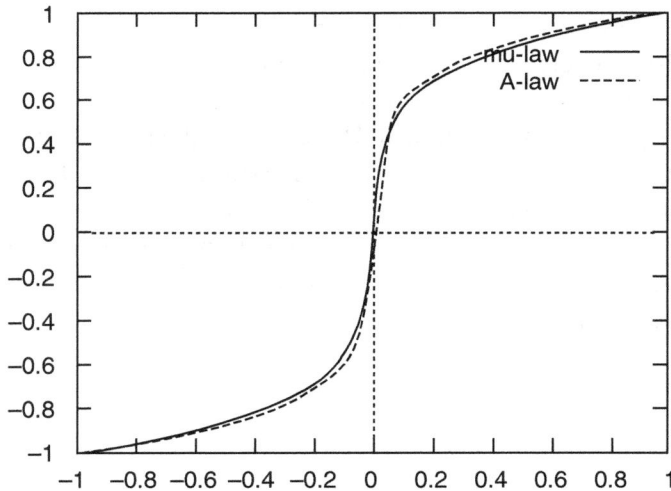

Figure 1.3: Logarithmic Quantization

Most implementations, however, do not use a linear quantization scheme (where the finite set of values to choose from is uniformly spaced) like PCM. Instead, a process known as companding is used. Companding (COMPression – expANDING) expands small values and compresses large values. In other words, when a signal goes through a compander, small amplitudes are mapped into a larger interval and larger amplitudes are mapped into a smaller interval. In this way, more quantization levels are used for the values that originated from small amplitudes (see Figure 1.3). This scheme is equivalent to applying nonuniform quantization to the original signal, where smaller quantization levels are used for smaller values and larger quantization levels are used for larger values.

The purpose of companding is to account for the fact that perceived intensity or loudness is not linear. We are more sensitive to sound at different volumes. With a strictly linear companding technique, the perceived change from, say, a value of 10 to 11 would be very different from the perceived change in a value of 250 to 251.

There are two standard forms of PCM: mu-law and A-law. Both attempt to compensate for this by using a logarithmic mapping and both produce 8-bit values every 125 µs, leading to a 64-kbps data stream.

Mu-law is popular in North America and Japan, and uses the following formula:

- $P = \ln(1 + uS) / \ln(1 + u)$ where S is the input sample, P is the output value and u is a constant with value. 255. In the formula "ln" refers to the natural logarithm function.

A-law is popular in Europe and uses the following formula:

- $P = a*S/ (1 + \ln a)$ for $S \le 1/a$ where a is a constant with value 87.6

- $P = (1 + \ln a*S) / (1 + \ln a)$ for $1/a \le S \le 1$

A-law is, in theory, easier for computers to implement. In either case, the result is a 64-kbps data stream consisting of 8-bit values produced every 125 µs. This stream is convenient for digital telephony to handle, and several communications standards have evolved to deal with such streams. One, known as T1, defines a protocol between two telephony devices where 24 digital voice streams (known as channels or time slots) can be transmitted over the same physical medium (wire or telephony "trunk"). T1 links operate at a speed of 1.544 mbps. The technology that allows multiple voice calls to share the same physical link through protocols such as T1 is referred to as multiplexing. Often in T1, one of the 24 channels is used to carry voice signaling instead of voice sampling. We discuss this in section 1.4.

So, to summarize, at some point in the path between caller A and caller B in today's PSTN, analog voice from the caller's handset will be digitized. In a PSTN, this usually takes place in the end office closest to your home.

1.3 PSTN Architecture

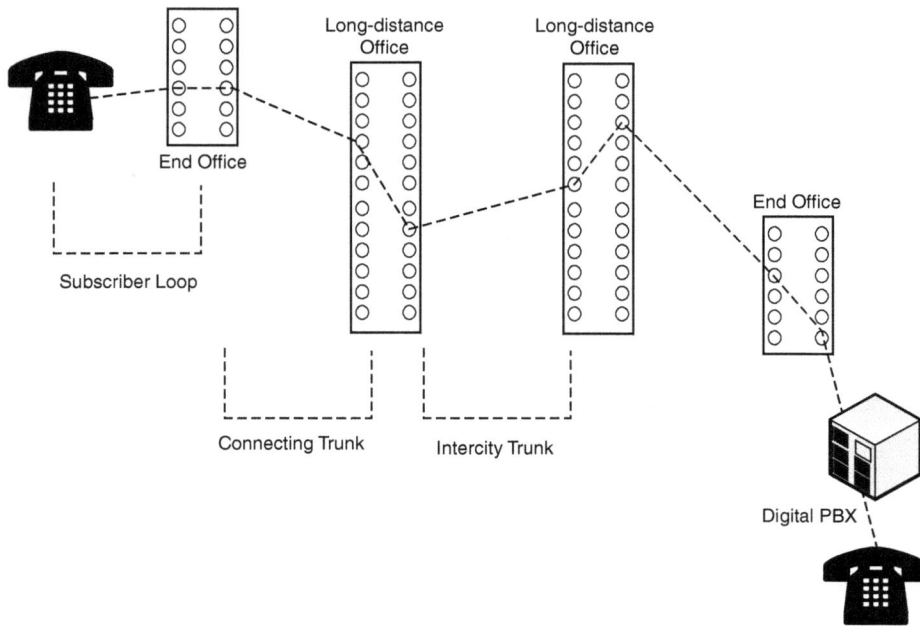

Figure 1.4: PSTN High-level Architecture

Figure 1.4 gives a high-level overview of the current PSTN architecture. The customer premises equipment (CPE) is typically a telephone. This connects via a dedicated pair of wires (often known as twisted pair) to the local office (aka central office). This part of the network that connects the end user to the local office is also known as the local loop, or the access network. Since many telephones (often in a single geographical area) connect to a central office, it is possible for calls made within a geographical area to be completed within the access network.

However, for calls destined to far-away geographical areas, long-distance offices (aka Class 4 switches) come into play. The local office is connected to long-distance offices via trunks, which can be thought of as huge capacity pipes. When a local office determines that the call is meant for a telephone not connected directly to it, it routes the call to the appropriate Class 4 switch. This Class 4 switch is then responsible for routing this call to the appropriate Class 5 switch, which in turn will route it to the end telephone. For international calls, another level of hierarchy comes into play, but the basic idea of hierarchical routing remains the same.

Thus far, we have discussed the PSTN architecture that carries voice calls, i.e., the media network. However, the PSTN really consists of two logically separate networks: the signaling network and the media network. To understand the difference between signaling and media, consider what happens when you pick up your telephone and make a call. You get a dial tone, dial digits, hear a ring-back tone and are then connected to the called party if (s)he answers the call. Notice that a whole lot of things happen before the voice actually starts flowing. Signaling refers to the overall process of going off hook, getting a dial tone, dialing digits, getting a ring back and finally getting a call connected. The media network comes into play only after the call is connected and is used for carrying the voice. These two logically separate networks are implemented as two physically separate networks in the PSTN. We have discussed the media network in this section and will discuss the signaling network in section 1.4.

To summarize, the media network consists of the physical wires (trunks) that carry voice calls and the switches that connect these trunks. It is the media network that reaches the end users at home. The end user's phone is connected to the local connection office aka the local telephone exchange aka central office (CO). These local telephone exchanges are connected to each other and to the tandem office by trunks. The trunks are used for carrying voice traffic between the switches and operating multiplexing protocols such as T1. The media network is therefore responsible for carrying voice traffic from one end user to another.

1.4 Signaling

1.4.1 Signaling in the Local Loop

As users of the PSTN, we exchange signaling with network elements all the time. Examples of signaling between a telephone user and the telephone network include: physically going on and off hook, ringing, dialing digits, providing dial tone, accessing a voice mailbox, sending a call-waiting tone, dialing *66 (to retry a busy number), etc.

Signaling in the local loop has been traditionally in-band—i.e., signaling takes place over the same path as the conversation. Basic signaling (e.g., signaling that a call needs to be placed or is waiting to be accepted) is done by changing the analog state of the local loop. For example, an incoming call is signaled by generating a cyclical ring voltage that in turn causes the ringer in the phone to turn on. More advanced signaling such as dial tone, dialed digits, and ringing tones are all audio signals that travel over the same channel on the same pair of wires in the local loop. When the call signaling is completed, voice is carried over the same path that was used for the signaling.

One question with in-band signaling is what happens to the analog signaling when the voice stream is digitized and converted to PCM. With the 64-kbps digital representation described above, how do we convey that the user has gone on hook or that the phone is ringing? The solution is to "borrow" some of the bits normally carrying voice samples and use them to carry signaling information instead. This is referred to as "robbed-bit" signaling and is used in digital trunks like T1 (where it is referred to as channel associated signaling or CAS). Robbed-bit

Name of Signal	Calling Station	Originating End Office	Intermediate Exchange(s)	Terminating End Office	Called Station
Connect					
Disconnect					
Answer (off-hook)					
Hang-up (on-hook)					
Delay-dial (delay pulsing)					
Wink-start					
Start dial (start pulsing)					
Dial tone					
Stop					
Go					
Called station identity					
DTMF pulsing					
Dial pulsing					
Multifrequency pulsing					
Calling station identity					
Verbal			Operator identification		
MF pulsed digits			Automatic identification		
Line busy					
Reorder					
No circuit					
Ringing					
Audible ringing					
Ringing start					
Recorder warning tone					
Announcements					

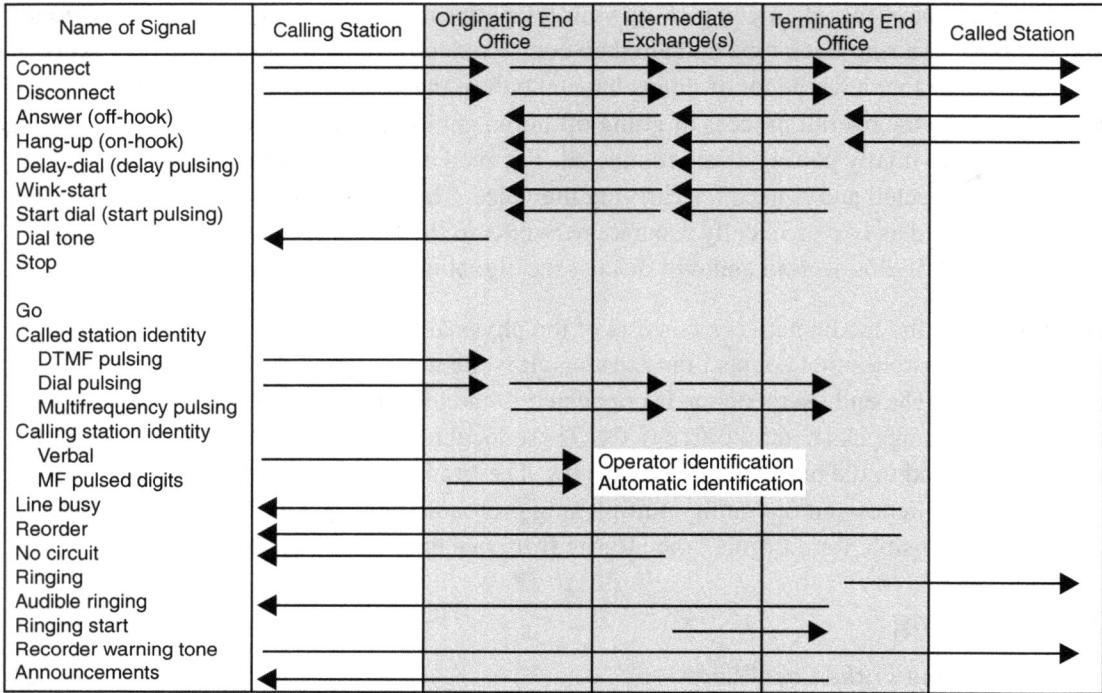

Note: A broken line indicates repetition of a signal at each office, whereas a solid line indicates direct transmittal through intermediate offices.

Figure 1.5: PSTN Signaling

signaling has a minimal impact on voice quality, as it works out that only one out of every 48 bits needs to be stolen.

Unlike in-band signaling, out-of-band signaling does not take place over the same path as the conversation. Instead, it establishes a separate digital channel for the exchange of signaling information. An example of this is the integrated services data network (ISDN). ISDN is an all-digital phone network where end user voice and signaling are converted to the digital domain in the customer premises (as opposed to being conveyed as analog over the local loop). When ISDN runs over T1 lines (known as the primary rate interface or PRI), it utilizes 23 out of the 24 timeslots for carrying voice (and possibly data). These are referred to as "B" channels, where "B" stands for bearer. One channel (channel 16) is dedicated to carrying voice-signaling information. This channel is referred to as the "D" channel. A lower-rate ISDN interface, the basic rate interface or BRI, uses one "D" channel with two "B" channels.

1.4.2 Signaling in the Network

Just like local-loop signaling, signaling in the network (i.e., between switches in the network) was initially in-band. Therefore, the signals to set up a call between one switch and another always took place over the same trunk that would eventually carry the call. Signaling took the form of a series of multifrequency (MF) tones, much like touch-tone dialing between switches.

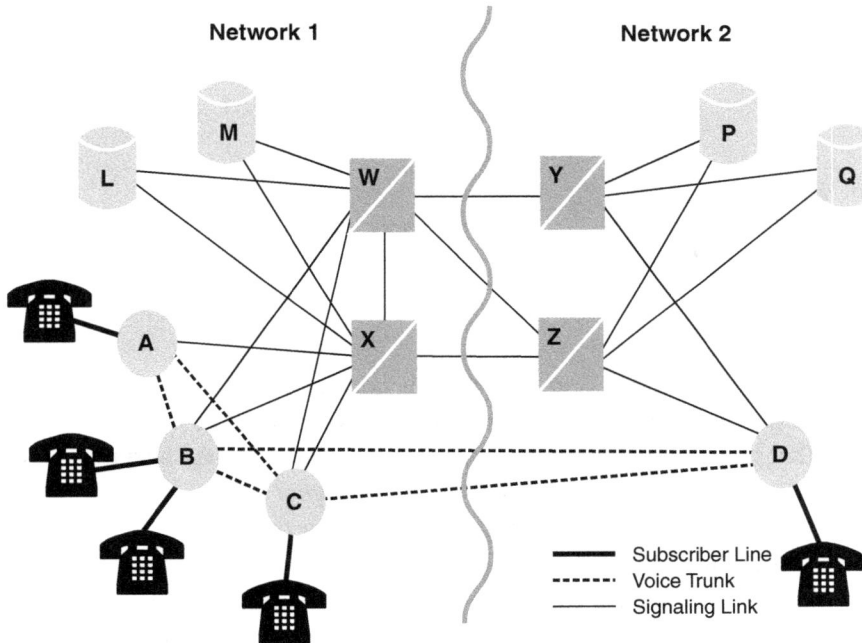

Figure 1.6: SS7 Architecture

However, this approach suffered from some limitations, which could be solved by using out-of-band signaling. Signaling links are used to carry all the necessary signaling messages between nodes. Thus, when a call is placed, the dialed digits, trunk selected, and other pertinent information are sent between switches using their signaling links, rather than the trunks which will ultimately carry the conversation. Out-of-band signaling has several advantages that make it more desirable than traditional in-band signaling.

- It allows for the transport of more data at higher speeds (56 kbps can carry data much faster than MF out-pulsing).

- It allows for signaling at any time in the entire duration of the call, not only at the beginning.

- It enables signaling to network elements to which there is no direct trunk connection.

1.4.3 SS7

The signaling network in the PSTN uses SS7 (Signaling System # 7) for call control. SS7 is an out-of-band (OOB) common-channel signaling (CCS) system. This means that the SS7 messages are carried on a logically separate network (out-of-band[3]) from the voice calls and that the signaling messages for all voice calls use this same network (common-channel). The SS7 network basically consists of signaling points (SP) exchanging control messages to perform call management.[4] There are primarily two types of signaling points: SSP and STP.

SSP	STP	SCP
Signaling Switching Point	Signaling Transfer Point	Signaling Control Point

Figure 1.7: SS7 Node Types

An SSP (service switching point) is an SP which is coresident with the local connection office (or the tandem office) and has the ability to control the voice circuits of its switch. The SSP is a logical entity that may be implemented in the switch itself or it may be a physically separate computer connected to (and controlling) the switch. In Figure 1.8, entities labeled 1 through 6 would each have an SSP associate with them. Note that, for the purpose of the signaling network, these entities are not connected directly to each other: the trunks that connect them are part of the media network used for carrying voice, not signaling messages. An STP (signaling transfer point) is an SP capable of routing call-control messages between the SSPs. An STP is therefore used by one SSP to route messages to another SSP. In Figure 1.8, entities labeled 7 and 8 are STPs.

To understand the overall picture, realize that since the PSTN media network is a connection-oriented network, the end-to-end connection between the calling party and the called party needs to be established before the call is "connected." This means that all switches in the media-path need to reserve resources (bandwidth, buffers, etc.) as part of signaling. This connection-oriented networking is known as circuit switching.

As an example, we go back to what happens when you pick up your phone and make a call.

---

[3] The nomenclature makes sense if you see the media network as the band carrying the voice.
[4] SS7 also specifies other nodes like an SCP used for advanced services, but those are irrelevant for the purposes of this discussion.

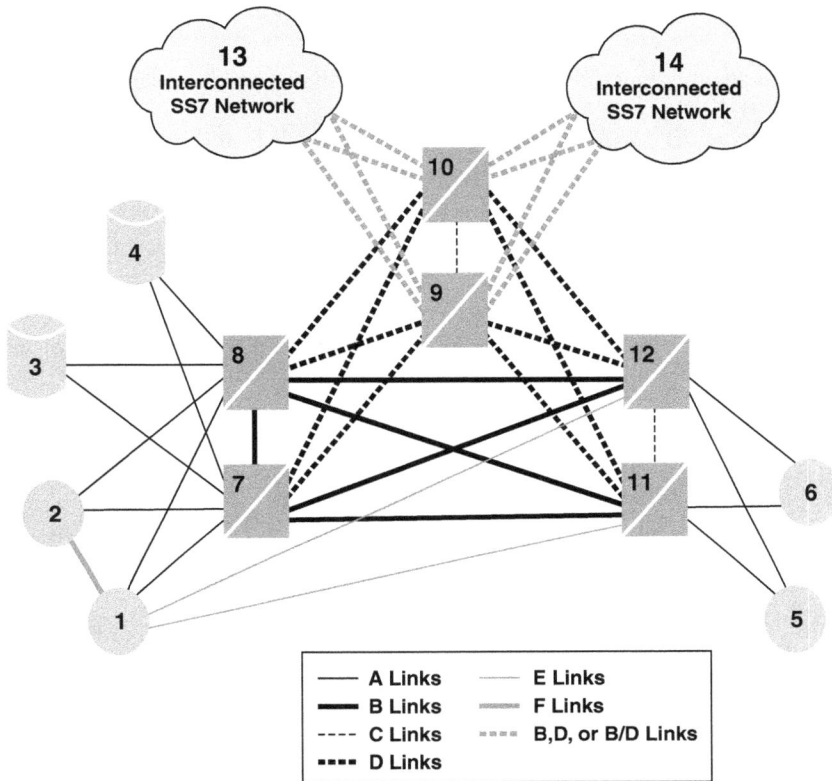

Figure 1.8: SS7 Nodes in Network Architecture

1.4.4 Call-Setup

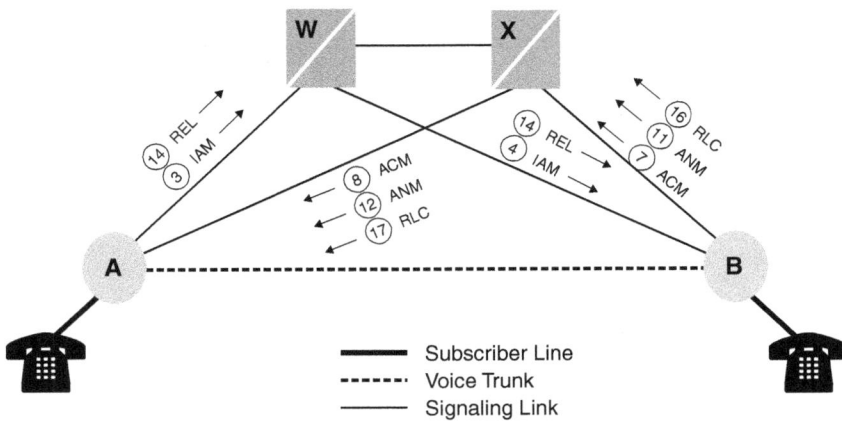

Figure 1.9: SS7 Signaling Messages for Call Setup

In this example, a subscriber on switch A places a call to a subscriber on switch B.

1. Switch A analyzes the dialed digits and determines that it needs to send the call to switch B.

2. Switch A selects an idle trunk between itself and switch B and formulates an initial address message (IAM), the basic message necessary to initiate a call. The IAM is addressed to switch B. It identifies the initiating switch (switch A), the destination switch (switch B), the trunk selected, the calling and called numbers, as well as other information beyond the scope of this example.

3. Switch A picks one of its A links (e.g., AW) and transmits the message over the link for routing to switch B.

4. STP W receives a message, inspects its routing label, and determines that it is to be routed to switch B. It transmits the message on link BW.

5. Switch B receives the message. On analyzing the message, it determines that it serves the called number and that the called number is idle.

6. Switch B formulates an address-complete message (ACM), which indicates that the IAM has reached its proper destination. The message identifies the recipient switch (A), the sending switch (B), and the selected trunk.

7. Switch B picks one of its A links (e.g., BX) and transmits the ACM over the link for routing to switch A. At the same time, it completes the call path in the backwards direction (towards switch A), sends a ringing tone over that trunk towards switch A, and rings the line of the called subscriber.

8. STP X receives the message, inspects its routing label, and determines that it is to be routed to switch A. It transmits the message on link AX.

9. On receiving the ACM, switch A connects the calling subscriber line to the selected trunk in the backwards direction (so that the caller can hear the ringing sent by switch B).

10. When the called subscriber picks up the phone, switch B formulates an answer message (ANM), identifying the intended recipient switch (A), the sending switch (B), and the selected trunk.

11. Switch B selects the same A link it used to transmit the ACM (link BX) and sends the ANM. By this time, the trunk also must be connected to the called line in both directions (to allow conversation).

12. STP X recognizes that the ANM is addressed to switch A and forwards it over link AX.

13. Switch A ensures that the calling subscriber is connected to the outgoing trunk (in both directions) and that conversation can take place.

14. If the calling subscriber hangs up first (following the conversation), switch A will generate a release message (REL) addressed to switch B, identifying the trunk associated with the call. It sends the message on link AW.

15. STP W receives the REL, determines that it is addressed to switch B, and forwards it using link WB.

16. Switch B receives the REL, disconnects the trunk from the subscriber line, returns the trunk to idle status, generates a release complete message (RLC) addressed back to switch A, and transmits it on link BX. The RLC identifies the trunk used to carry the call.

17. STP X receives the RLC, determines that it is addressed to switch A, and forwards it over link AX.

18. On receiving the RLC, switch A idles the identified trunk.

1.5 Voice and Wireless Networks

Thus far we have talked about wired networks being used to carry voice. In this section we give a brief overview of how wireless networks are used to carry voice.

1.5.1 First-Generation Wireless Networks

The earliest wireless voice networks were deployed in 1980 and 1981 in Japan and Scandinavia. In the following years, various cellular systems were developed and deployed all over the world. Together these came to be known as the first-generation cellular systems. Even though these standards were mutually incompatible, they shared many common characteristics. The most prominent among them was that voice was transmitted by means of frequency modulation; that is, the air-interface in these standards was analog.

One of the first-generation wireless cellular systems was the advanced mobile phone system (AMPS) in North America. Figure 1.10 shows the prominent network components in the AMPS architecture. The *mobile station* (MS) is the end user terminal that communicates over the wireless medium with the *land station* (LS). The land station (also known as base transceiver station) is connected by land lines[5] to the *mobile telephone switching office* (MTSO). This was the AMPS architecture. The deployment of an AMPS wireless network required the deployment of MTSOs, LSs and the end user mobile stations.

---

[5] Land lines may physically be copper wires, optical fibers or microwave links.

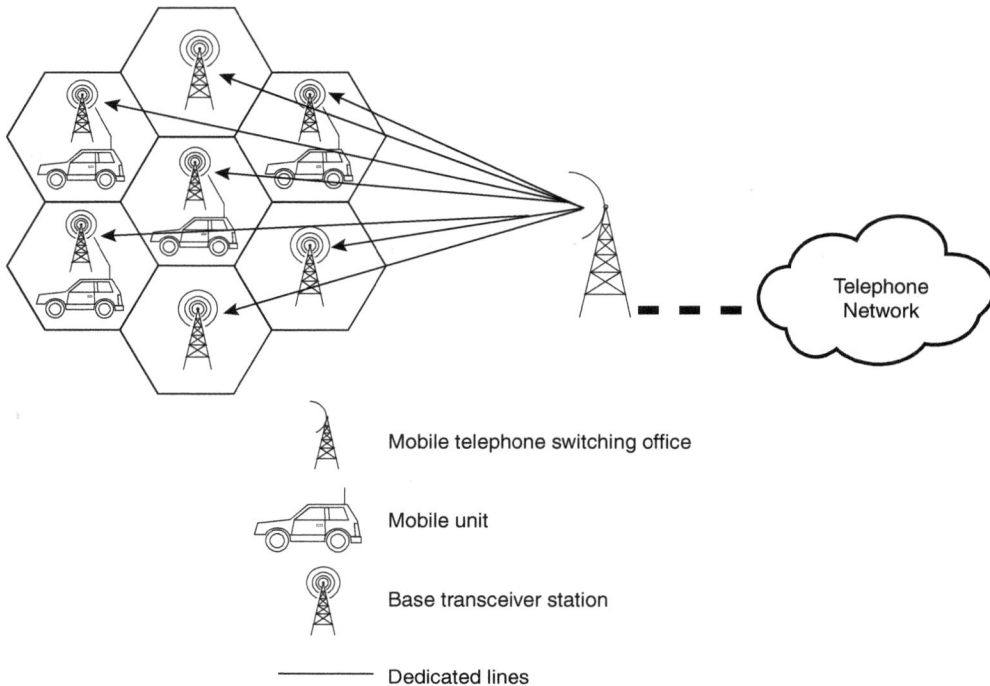

Figure 1.10: AMPS Architecture

When the mobile user dials a phone number, this phone number is relayed from the LS to the MTSO. The MTSO is basically a CO enhanced to support mobility in the wireless medium. Just like its wired counterpart, the MTSO consists of a switch connected to the media network of the PSTN and an SSP connected to the SS7 network. When the MTSO gets the called-party number, it uses the same procedure as any another CO to route the call. The PSTN is not aware that the end user is a wireless user and it sees the MTSO as just another CO. This makes routing calls between the MTSO and the PSTN easy. Once the call reaches the MTSO, it is the MTSO's responsibility to route the call to the end user's phone. It can do this because it uses location management to find or know the location of a MS at any given time. This is how calls get routed in a wireless network.

1.5.2 Second-Generation Wireless Networks

The first-generation wireless cellular networks specified the communication interface between the mobile station and the land-station; that is, it specified the air-interface but not the communication interface between the LS and the MTSO. This had far-reaching implications on the system architecture, in that the LS and the MTSO had to come from the same vendor, since the communication protocol between the LS and the MTSO was proprietary. The lack of coordination between various vendor switches meant that, even though subscribers could make and receive calls within the areas served by their service provider, roaming services between service providers were spotty and inconsistent.

Even though the wireless industry in the United States developed Interim Standard 41 (IS41) to address the roaming problem in first-generation networks by standardizing the communication protocol between the MTSOs, the problem still existed in Europe where there were as many as five mutually incompatible air-interface standards in different countries in Europe. This, at a time when Europe was moving towards a model of European economic integration, led the Conference of European Postal and Telecommunication (CEPT) to undertake the development of a continental (read pan-European) standard for mobile communication. This led to the global systems for mobile-communications (GSM) specification, with one of the primary underlying goals being seamless roaming between different service providers.

The term "second-generation cellular networks" is a generic term referring to a range of digital cellular technologies. Unlike the first-generation networks, all second-generation networks have a digital air interface. With an estimated 1 billion subscribers all over the world, the most dominant second-generation technology is GSM. GSM has several salient features. It combines time division multiple access (TDMA) and frequency division multiple access (FDMA) to specify a hybrid digital air interface. Therefore, unlike AMPS, where a logical channel could be specified by specifying just the carrier frequency, a logical channel in a GSM needs to be specified using a carrier frequency (FDMA) and a timeslot (TDMA). Another important feature of GSM is that it specifies not only the air interface but many other interfaces in the GSM network, as shown in Figure 1.11.

The end user equipment (typically a cell phone) is known as the mobile equipment (ME) or the mobile station (MS). The term MS refers together to the physical device, the radio transceiver, the digital signal processors, and the subscriber identity module (SIM). The SIM is one of the great ideas to come out of the GSM standard. It is a small electronic card that contains user-specific information like the subscriber identity number, the networks that the user is authorized to use, the user encryption keys and so on. The concept of separating the subscriber-specific information from the physical equipment (the phone) allows the user to use their service from a variety of equipment, if they desire.

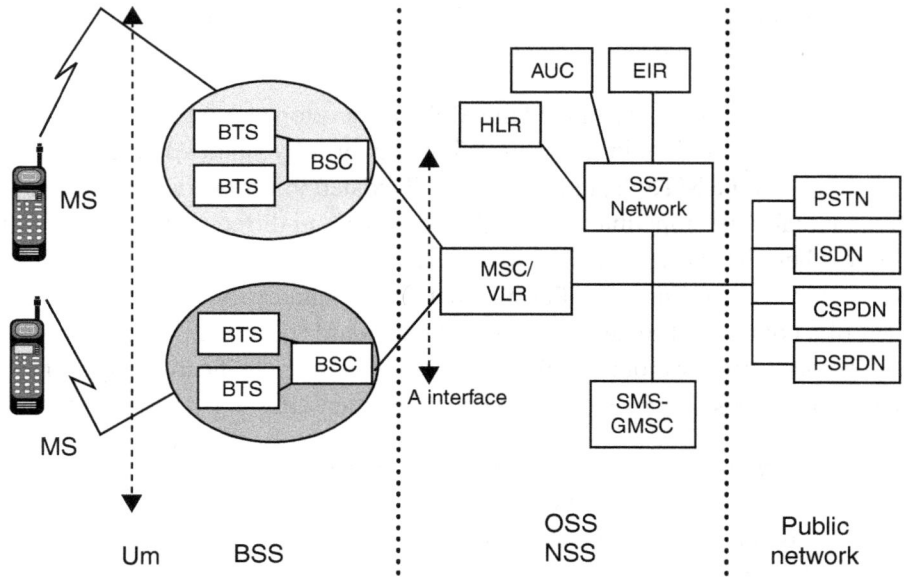

Figure 1.11: GSM Nodes and Interfaces

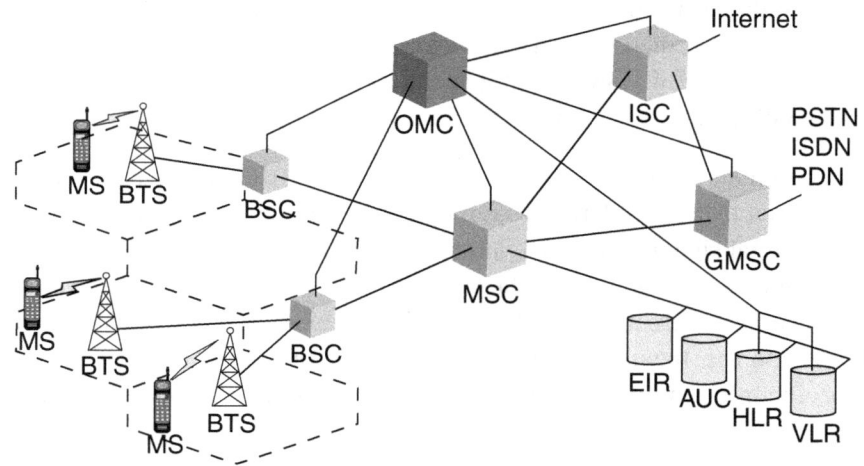

Figure 1.12: GSM Network Architecture

The mobile equipment communicates with the base transceiver station (BTS), which consists of a radio transmitter and a radio receiver and is the radio termination interface for all calls. The interface between the MS and BTS is known as the Um interface. The BTS is the hardware that defines the cell (in that each cell has exactly one BTS). It consists of a radio antenna, a radio transceiver and a link to the base station controller (BSC), but it has no intelligence. The intelligence (software) that controls the radio interface sits in the BSC and is responsible for things like channel and frequency allocation, tracking radio measurements, handovers, paging, and so on. Each BSC usually controls multiple BTSs and the interface between these two components is known as the Abis interface. The BSC and the BTSs together constitute the base station subsystem (BSS) of the GSM network. Beyond the BSS exists the GSM core network.

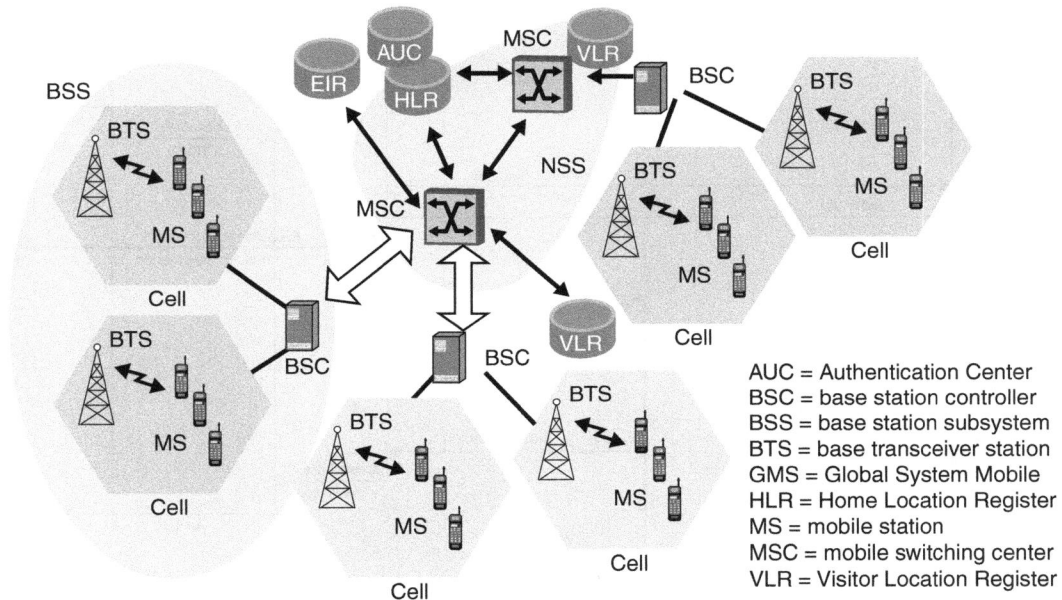

AUC = Authentication Center
BSC = base station controller
BSS = base station subsystem
BTS = base transceiver station
GMS = Global System Mobile
HLR = Home Location Register
MS = mobile station
MSC = mobile switching center
VLR = Visitor Location Register

Figure 1.13: GSM Network

The GSM network components described so far are probably sufficient to provide the most basic wireless voice service. However, there are four important databases in the GSM core network that allow the GSM network to provide seamless service to the end user. The first is the home location register (HLR), which stores information on each subscriber that "belongs" to it. This includes information like the subscriber's address, billing information, service contract details, and so forth. The HLR is therefore the central repository of all information regarding the user.

The visitor location register (VLR) is a database in the GSM network that is required to achieve seamless roaming in all service areas in the network. Unlike the HLR, which is usually unique at the service provider level, the VLR is one per MSC and keeps track of all users currently in the area being served by this MSC. To understand the need for a VLR, consider what happens when a call from the PSTN needs to be terminated on a mobile phone. The PSTN will route the call to the GMSC of the service provider to which the terminating phone-number belongs. The GMSC then queries the HLR regarding this user. The HLR contains a pointer to (the address of) the VLR where the subscriber is currently located. The GMSC can therefore route the call to the corresponding MSC, which would then terminate the call on to the mobile equipment.

The magic of how the HLR knows the current VLR is a complex procedure of location updates, as explained in Figure 1.14.

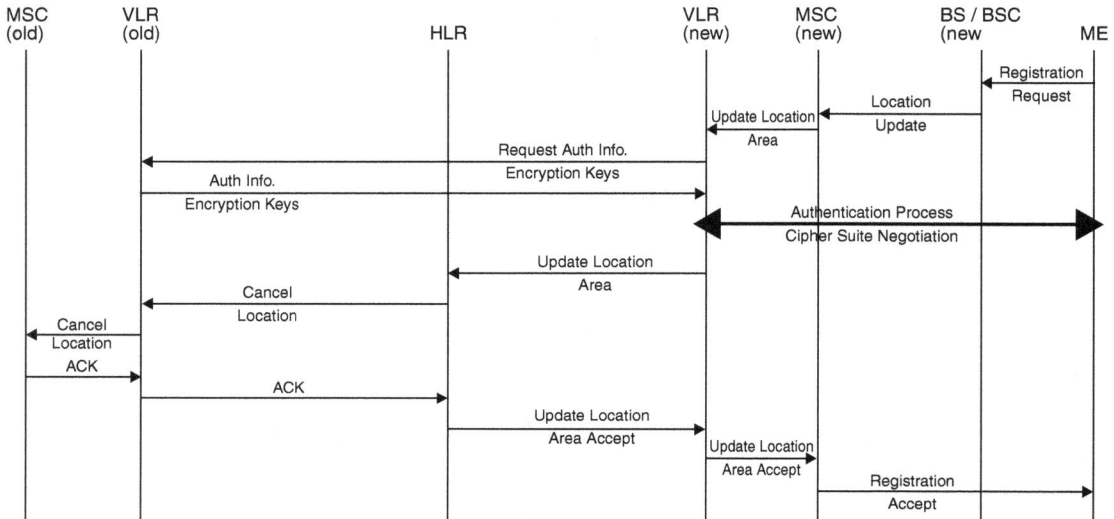

Figure 1.14: Handling Mobility in GSM

Whenever mobile equipment detects that the signal from its current BTS is too low (below a certain threshold), it starts the roaming procedure to connect to the BTS with the strongest signal strength. To do this, the mobile equipment sends a registration request to the new BTS. In turn, BTS sends a location update to its MSC. The MSC then updates its VLR to update information regarding this user. This VLR now contacts the old VLR where the ME was previously registered to get the authentication and encryption keys for this user. Also, the VLR contacts the ME's HLR to update the information regarding this ME. It is the HLR which in turn updates the old VLR to remove the subscriber's identity.

The detail and complexity of the GSM standard can be estimated by the fact that the total length of the standard is more than 5000 pages long. The fact that the interface between each network component in GSM is specified allows service providers to purchase different network components from different vendors. Note, however, that the only interface GSM specifies at the physical layer is the air interface between the MS and the BTS. All other interfaces are specified from Layer 2 above, leaving the physical layer implementation to the service provider; for example, the service provider may decide to have the physical interface between the BTS and the BSC as a microwave link or as a fiber-optic link, depending on the requirements.

With this background, we now look at how a call originating from the PSTN destined to a GSM subscriber proceeds:

1. Call-setup messages reach the GMSC through the PSTN.

2. The GMSC contains a table linking MSISDNs to their corresponding HLR. It uses this table to interrogate the called subscriber's HLR for the MSRN of the called subscriber.

3. The HLR typically stores only the SS7 address of the subscriber's current VLR, and does not have the MSRN. The HLR therefore queries the subscriber's current VLR.

4. This VLR will temporarily allocate an MSRN from its pool for this call and inform the querying HLR of the MSRN.

5. The HLR forwards this MSRN to the GMSC.

6. The GMSC uses this MSRN to route the call to the appropriate MSC.

7. When the appropriate MSC receives the call request, it looks up the IMSI corresponding to the MSRN in the call request and then broadcasts a page in the current Location Area of the subscriber.

8. The appropriate ME responds to the paging request.

Similarly, a call originating from the GSM subscriber destined to the PSTN proceeds as follows:

1. When the user presses the "send" button on their phone, the MS sends the dialed number to the BTS.

2. The BTS relays the dialed number to the MSC.

3. The MSC first checks to see if this number belongs to one of its own subscribers who may be reached "locally" without accessing the PSTN. The MSC can find this out by referring to its HLR.

4. If the called party is a subscriber, the MSC can also determine its current location using the HLR and then forward the call to the appropriate MSC/VLR.

5. If, however, the called party is not a subscriber, the MSC uses the PSTN to route the call.

6. Once the MSC receives an acknowledgment from the remote CO, the MSC tells the BTS to allocate voice channels to the MS for this call.

7. The BTS allocates voice channels for the MS and informs the MS about these.

8. The MS can now start the voice conversation using the voice channels allocated to it.

1.5.3 Third-Generation Wireless Networks

As of the writing of this book, 2G networks like GSM continue to be the most widely deployed wireless networks. It is estimated that GSM alone has over 1 billion subscribers in this world. There is no question that the 2G networks have been hugely successful in satisfying the needs of customers today. So, what is the need for next-generation wireless (3G) networks? That, in fact, is the billion-dollar question. Most wireless service providers in the United States and Europe spent billions of dollars in buying the radio spectrum in which 3G was to operate. The expectation at that time was that the bandwidth capacity and the features provided by 3G would soon be needed for next-generation wireless applications.

As it turns out, there have been no new "killer applications" that would justify moving from 2G to 3G. This, combined with the success of add-on technologies, such as general packet radio service (GPRS) and so on, have allowed 2.5G (2G + add-on services) to serve the demands of subscribers to date. This has left the service providers who bought the 3G spectrum in a tight spot. Nobody knows for sure what's next for the wireless industry after 2.5G. There are basically two camps—one camp believes that 3G will happen soon enough, if only because of the huge amounts of money invested in the technology. This camp is supported by the fact that early deployments of 3G are beginning to appear in Japan. The other camp believes that there will soon be an IP-based 4G wireless technology ready to give 3G a huge blow. This camp is supported by the fact that a lot of service providers are holding off 3G deployment, waiting for subscriber demand or 4G. The thing about 4G, however, is that nobody knows for sure what shape and form it will take. So whether 3G will happen or not is still an open question.

In any case, we will take a look at 3G in this section. Figure 1.15 shows the 3G network architecture. The architecture looks very similar to the 2G network but there are important differences.

First, the radio interface uses code division multiple access (CDMA). Even though this is not something new in 3G, it shows the acceptance of CDMA as a superior technology as compared to TDMA in GSM. Second, the network emphasizes the movement of intelligence in the network from the core of the network to the periphery. Third, the architecture shows the integration of the voice network and the IP network.

The universal mobile telecommunications system (UMTS) network architecture is pretty similar to the 2G network architecture, with subtle differences. The 3G equivalent of the BTS is known as Node-B. It differs from the BTS primarily in that the air-interface used by 3G is

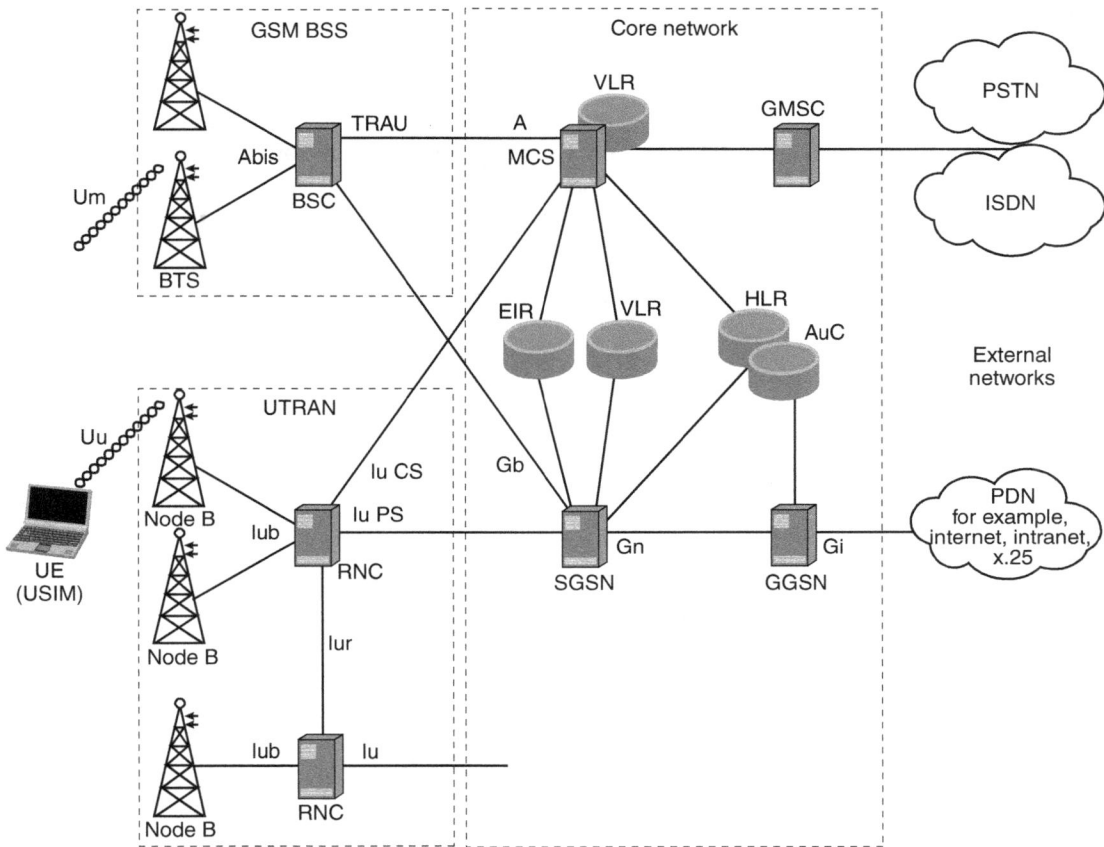

Figure 1.15: 3G Network Architecture

always CDMA, unlike 2G where there were multiple standards that could be used for the air interface. The 3G equivalent of the BSC is the radio network controller (RNC). It differs from the BSC in that it takes on many more responsibilities. Most importantly, the role of mobility management is moved from the core of the network to the RNC (effectively the access network—that is, the edge of the network). The RNC connects to the same core network that is used by 2G networks. Therefore, the 3G standard is different from the 2G standard only on the access side: the core network is kept the same.

Since the core network is kept the same as that of the 2G networks, call routing is almost the same as in 2G networks. The differences in the call-setup procedure are minimal and are mostly related to who does what, since more responsibilities have moved to the edge of the network. We do not therefore repeat the connection-setup procedure, since for the purposes of this text this is the same as connection setup in 2G networks.

1.6 Summary

This chapter has presented some background on today's telephony networks. To summarize, these networks are connection oriented. When I place a call to someone, the phone system locks up a set of resources (e.g., timeslots on T1 trunks) for the duration of the call. This is known as circuit switching and provides a good way to ensure consistent user quality of service during the call. Two types of signaling methods are used: in-band and out-of-band. In-band signaling is almost always used on the local loop, but out-of-band signaling is the norm within the core of the modern phone network. Phone networks are hierarchical in nature.

The Data World

2.1 Introduction

Having discussed traditional voice networks in the previous chapter, will now turn our attention to data networks. We will concentrate on packet-based data networks—in particular the most familiar one to most of us, the Internet. This chapter will be no more than a brief introduction to the topic. For more information, the reader is directed to the references listed at the end of this book.

2.2 Brief History

Early computer data networks simply used the existing voice network to carry data traffic. As we saw in Chapter 1, these voice networks were circuit switch based, meaning that network resources needed to be permanently dedicated between nodes wishing to pass data. These resources had to be dedicated even if the nodes were not actively communicating, such as during silence periods during an ongoing conversation. Furthermore, unused capacity between nodes A and B could not easily be reassigned to the connection between node A and other nodes.

In 1961, Leonard Kleinrock, then at MIT, presented an alternative approach in his Ph.D. thesis proposal (*Information Flow in Communication Nets*, May 1961). In this thesis, Kleinrock proposed to study the behavior of data networks consisting of multiple nodes and communication links. In these data networks, information is broken into smaller pieces (packets or messages) to be transmitted and then reassembled at the destination. On the way to the destination, these packets will visit one or more intermediate nodes and links in the network. These ideas, inspired in part by queuing theory, formed the core concept of what is known as packet switching. Intermediate network nodes are referred to as packet switches, or in TCP/IP networks as *routers*.

The ideas of packet switching were picked up by the Defense Department Special Projects Agency (DARPA), when researchers realized that another benefit of the technology was survivability. In such networks, the loss of a percentage of the communication nodes and links can be handled by rerouting the individual packets through nodes and links that are still operating.[1] Contrast this to the hierarchical phone network, where the loss of a Class 5 switch will cause a total loss of service.

By 1969, a small (i.e., four node) experimental packet data network was operational. This formed the start of a network known as the Arpanet, which was used for many years as a Defense Department research network. It provided a test bed for developing packet communication protocols, including the ubiquitous Transmission Control/Internet Protocol (TCP/IP), which we will discuss below.

In 1983, the Arpanet was split into two parts, a secure DOD network and an unclassified research network. This latter network, under various names and transformations, became what we now know as the Internet (the Arpanet itself as a government-managed research network was shut down in 1990). The number of devices on the network has been exponentially increasing. For example, in 1989 the number reached 100,000, while five years later the number reached 5 million. The number of devices today (January 2006) is estimated to be over 390 million (note: this includes more than just hosts or computing servers; for example, Voice-over-IP terminals are included in this count).

2.3 The OSI Seven-Layer Model

To describe packet data networks, it is traditional to begin with a description of the Open Systems Interconnection (OSI) communication protocol model. This model is illustrated in Figure 2.1, and divides communication into seven layers.

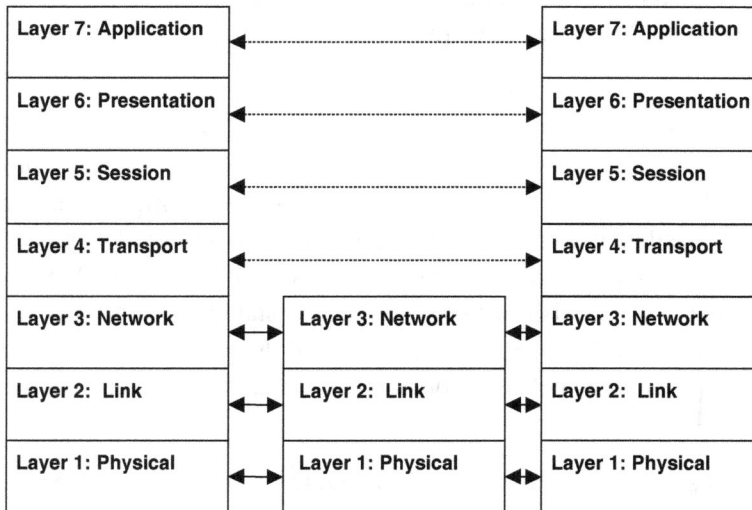

Figure 2.1: The Seven-Layer ISO Model

---

[1] Hence the attraction to the Defense Department.

Layer 1 or *physical layer* defines the mechanisms used to physically transmit and receive raw digital data. The classical example of a physical layer is the 802.3 Ethernet protocol. This protocol defines how devices use Ethernet physical cabling to communicate. If one cares to examine a physical-layer protocol specification (such as 802.11b), he would see information regarding the modulation techniques to be used, data ordering, error correction and transmission rates.

Layer 2 defines the *link layer*, which is responsible for how devices communicate over the physical layer. Examples of link-layer protocol concerns include how data is to be packaged and presented to the underlying physical layer, what kind of acknowledgment and control messages are required, and how physical-layer error conditions are to be handled. A key component of a link-layer protocol is the media access subprotocol. This subprotocol defines how devices access and share the physical layer. In the case of 802.3, the standard defines the classic Ethernet Carrier Sense/Multiple Access with Collision Detection (CSMA/CD) protocol. We will go into detail on the 802.11 link layer in Chapter 4. The link layer also defines how devices are addressed. For 802.x link-layer protocols, this address is called a MAC (or media access address) and is assigned by the device manufacturer.

Link-layer protocols can be classified in various ways, including (a) connection-oriented or connectionless and (b) reliable or unreliable. An example of a reliable, connection-oriented link-layer protocol is HDLC. An example of an unreliable, connectionless link-layer protocol is 802.3 (Ethernet).

Layer 3 is used to represent the network layer. This layer is responsible for end-to-end communication across multiple layer 1 and layer 2 networks. A key component of layer 3 protocols is routing—how the path from transmitter A to receiver B is figured out and maintained. Like layer 2, layer 3 protocols can be categorized as connection oriented or connectionless. A connection-oriented layer 3 protocol (e.g., ATM) will set up what is referred to as a virtual circuit across the network, whereas a connectionless layer 3 protocol (e.g., IP) will route each packet individually based on the information contained in the packet (think of this as a postcard in the postal network).

The transport layer (layer 4), is the first layer in the model that is truly between the actual endpoints in the data transfer. The function of the transport layer is to accept data from the upper layers (e.g., session), break this into smaller units for transfer, and provide it to the network layer. On the receive side, the transport layer is responsible for receiving the smaller data units, ordering them correctly and reassembling them into something that the upper layers can accept. The transport layer is also concerned with congestion control: the process of making sure that too much data is not flooding the network at any given instant of time.

Layer 5 is known as the session layer and presents the user (i.e., application and presentation layers) a standard interface to the network. The session layer is concerned with

25

session management, such as how the upper layers gain access to the network, how they are authenticated, how they discover other users, and how a communication session is set up and managed. Often the session layer will include transaction management components.

Layer 6 is the presentation layer. This layer is concerned with how the data to be transferred is to be represented. Different computer systems (software or hardware) will represent data each in their own native formats. Presentation-layer protocols define how data to be transferred needs to be transformed into a common format. Other presentation-layer concerns are data compression techniques and encryption. Examples of presentation-layer protocols include compression schemes, abstract syntax models (such ASN.1), and the HyperText Markup Language (HTML).

Finally, layer 7 is the application layer. This is the highest layer in the model, and is user (application defined). Examples of application-layer protocols are the File Transport Protocol (FTP), the TCP/IP terminal emulation program (TELNET), and the Simple Mail Transport Protocol (SMTP).

The seven-layer model is exactly this, a model. It does not necessarily apply totally to every network. For example, Figure 2.2 shows how the seven-layer model matches up to the industry standard TCP/IP protocol. As can be seen, not all layers are represented, and some layers are combined together.

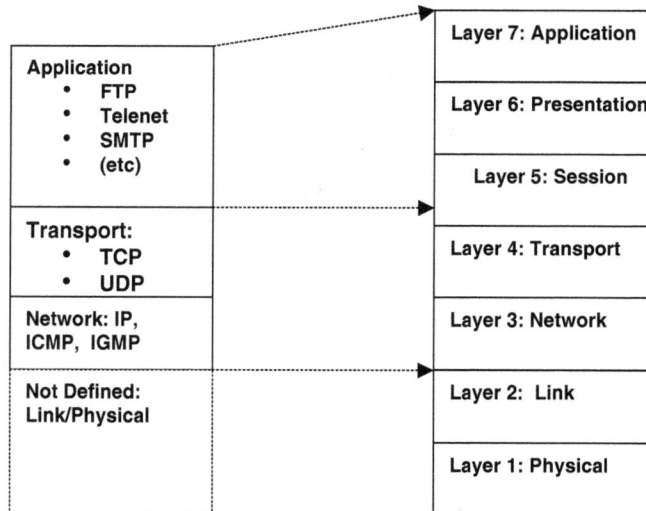

Application • FTP • Telenet • SMTP • (etc)	Layer 7: Application
	Layer 6: Presentation
	Layer 5: Session
Transport: • TCP • UDP	Layer 4: Transport
Network: IP, ICMP, IGMP	Layer 3: Network
Not Defined: Link/Physical	Layer 2: Link
	Layer 1: Physical

Figure 2.2: TCP/IP Protocol Suite Mapping to the Seven-Layer ISO Model

Given our theoretical model, we will next discuss the protocols used in the Internet, namely those listed on the right side of Figure 2.2. While these are commonly referred to as TCP/IP, they include more than just these two individual protocols. We will concentrate mainly at level 3 and above in this discussion. As is illustrated in Figure 2.2, the TCP/IP protocol suite does not define the link and physical layers and can, in fact, run over many different link- and physical-layer protocols. We will defer presentation of the link/physical layer protocols that are relevant to this book (i.e., 802.11) until Chapter 4.

The basic model for TCP/IP networks is shown in Figure 2.3. In the model, TCP/IP networks consist of end devices and routers. End devices attach them to a local network segment. Routers connect segments together.

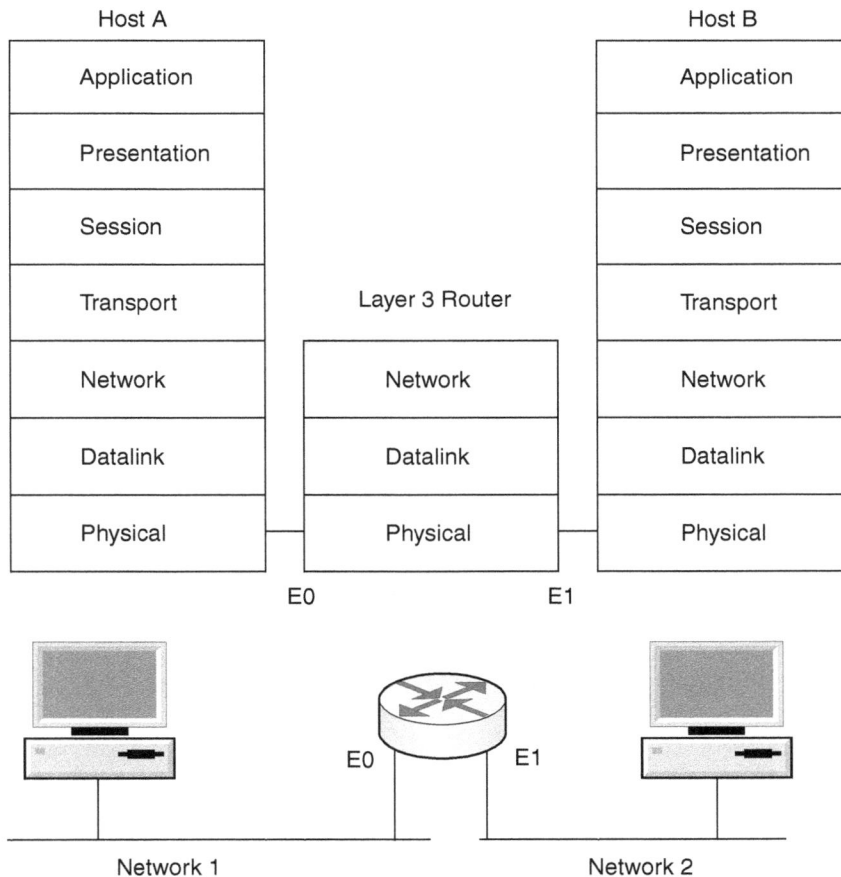

Figure 2.3: TCP/IP Network Model

2.4 The IP Protocol

The network layer component of TCP/IP is the Internet Protocol (IP). This protocol forms the backbone of the Internet. Every data packet (and control/management packet) that is *routed* through the Internet uses this protocol. The IP is defined in RFC791.

IP is interesting in that it provides *only* an insecure, unreliable, and connectionless data-delivery service through a network of IP devices. That is, a chunk of data to be sent over IP to a destination is not secured, is not guaranteed to arrive at the destination, and may take a different route through the IP network each time it is sent.

Each chunk of data to be sent over IP is combined with an IP header (Figure 2.3). The resulting packet is referred to as an IP *datagram*. This header is a minimum of 20 bytes long and includes the fields defined Table 2.1 below (for Version 4 of the protocol).

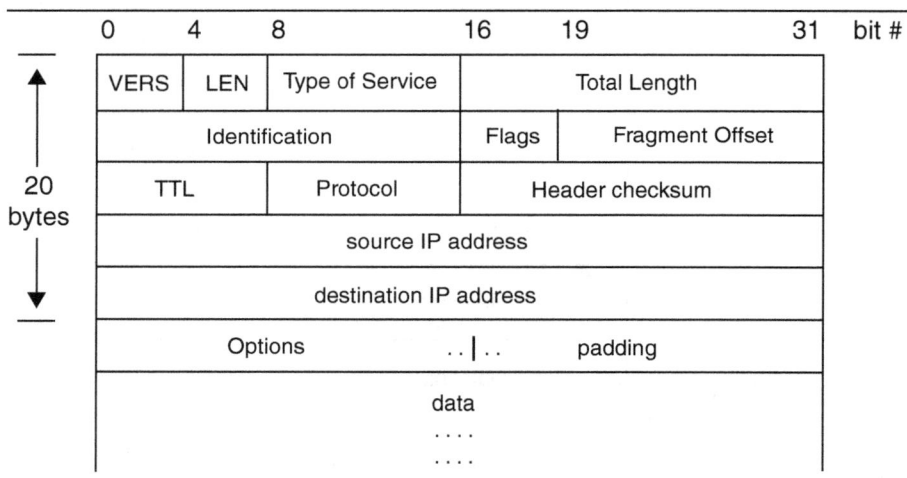

Figure 2.4: IP Datagram

Table 2.1: IP Version 4 Header

IP Header Field	Size (bits)	Purpose
Version	4	Identifies the version of the protocol. Version 4 (V4) is the typical standard, although version 6 of the protocol has been defined and is in use in some networks.
Header length	4	This gives the number of 32-bit words in the header. Typically set to 5 for IP V4 (i.e., a normal IP V4 header is 20 bytes long).
Type of service	8	Can be used to define a quality of service level for this datagram. For example, IP packets containing voice samples can be marked as high priority here.
Total packet length	16	Defines the total IP datagram size (in bytes and including the IP header). Note that 16-bit size here implies that largest possible IP datagram is 65535 bytes.
ID	16	Identifies each datagram. Typically incremented by 1 for each datagram sent by a particular source. Used for fragmentation/reassembly operations.
Fragmentation flags	3	These are used for fragmentation/reassembly operations.
Fragmentation offset	13	Used in conjunction with fragmentation/reassembly operations.
Time to live	8	This field sets an upper limit on the number of hops (intermediate IP devices) that the packet can visit on its way to the destination. Each device that sees the IP datagram decrements this field.
Protocol	8	This field identifies the next level protocol that is being carried in the datagram. Each IP transport protocol must be registered to receive an "official" protocol ID. For example, a value of 6 is reserved for TCP; a value of 17 is used for UDP.
Header checksum	16	This checksum is calculated over the IP header only, and is used to verify that the header has not been corrupted. Note that it is not over the entire datagram.
Source address	32	This is the address of the device generating the datagram.
Destination address	32	This is the address of the end-destination of the datagram and is not necessarily the address of the next hop for this datagram.
{Options}		Not commonly used. Various uses including: • Route recording • Loose/strict routing lists • Security handling restrictions

IP V4 uses a 32-bit number to define device addresses. This number is represented in human-readable form by the familiar "dotted decimal" notation (e.g., 192.168.1.1). This address is hierarchical in the sense that a certain number of bits are used to define a network, and the remaining bits are used to identify the device on that network. Early uses of IP addressing defined static classes of addresses; for example, an address in the range 192.00.00.00 to 223.255.255.255 was referred to as a class C device where 21 bits were used for the network portion and 8 bits for the device portion. Similarly, there were class A, B and D addresses distinguished by the number of bits used for specifying the network portion of the address. However, this is not strictly true today. More generally, IP devices are told, during the configuration process, the partitioning (i.e., how many bits refer to the network address and how many refer to the device address) that is used in the network. This is done using a netmask: a netmask of 255.255.255.0 means that 24 bits are used for network addressing and 8 bits are used for device addressing, thus allowing this LAN to have a maximum of 128 devices. Similarly, a netmask of 255.255.254.0 means that 23 bits are used for network addressing and 9 bits are used for device addressing, thus allowing this LAN to have a maximum of 256 devices.

Most IP addresses are unicast, meaning that they identify a single endpoint. A subset of the address space is reserved, however, for broadcast or multicast addresses. With IP multicasting, IP hosts can join a multicast group (identified by its multicast IP address) and then receive broadcast IP datagrams. An IP protocol known as the Internet Group Management Protocol (IGMP) is used to manage multicast groups (i.e., joining and leaving) and the routing of these multicast IP datagrams. IP multicasting has use in applications such as streaming video (for example, IP TV) and a telephony service known as push-to-talk.

Note that IP addresses are network-level addresses. To actually reach a device on a network, the IP address needs to be mapped to a link layer (or MAC address in the case of 802.x networks). There is a standard protocol to facilitate this known as the Address Resolution Protocol (ARP); it works as follows:

- Each IP device maintains a cache of IP address to MAC address mappings that it is aware of, for each network it is attached to.

- When a new IP datagram arrives to be transmitted to a particular network, it examines this cache. If the mapping is present, then the device knows what link-level address to use to route the packet.

- If it is not present, the device sends a special link-level packet known as an ARP request. This packet is sent to all devices on the network segment (e.g., through a broadcast mechanism if this is supported by the link-level network), and is essentially asking for the owner of IP address X to speak up.

- The device having this address sends an ARP reply. The requestor uses the information in the message to populate the cache and the IP datagram can now be sent.

As mentioned above, IP does not guarantee reliability or provide a connection through the network. So what does it do exactly? The IP protocol provides the following services:

- Routing: IP datagrams can be routed through an IP network to the destination. Companion protocols such as ICMP and RIP (which are based on IP also) are used to build and maintain routing tables in intermediate IP routers. By examining the destination address header field of an IP datagram, routers are quickly able to figure out where to forward the datagram so that it ultimately will arrive at the destination.

- Fragmentation: As stated in the table above, an IP datagram may be as large as 65535 bytes. However, during the datagram's voyage to the destination it may encounter a link-layer protocol that supports a smaller maximum packet size. In this case, the IP datagram may be fragmented into smaller pieces to be able to traverse this link. The fragmentation fields in the IP header (ID, flags and offset) are used to carry enough information for the destination to reassemble the fragments. IP fragments are still IP datagrams, so they will be routed through the IP network.

- Multicasting and broadcasting: The IP address of FF:FF:FF:FF is used to define a broadcast address. A broadcast IP datagram is forwarded across all links. A separate IP address range is reserved for multicast services, known as multicast groups. A multicast IP datagram belonging to group X is forwarded only across links connecting to devices that have registered their membership to this group.

- Identification of the next protocol: The protocol field of the IP header provides an easy way to determine what protocol is being carried in the IP datagram.

- Datagram "self-destruct": An IP datagram that gets misrouted will eventually have its time-to-live field reduced to zero since each router reduces the value of TTL field by 1. This will cause the next router to receive the datagram to drop it, and thus IP datagrams are guaranteed to have a finite lifetime. This prevents loops, spirals, etc. in the network.

Before leaving IP, we should mention its companion protocol: the Internet Control Message Protocol (or ICMP). ICMP is a utility protocol that provides several services. It is carried in IP datagrams as well (and is identified by an IP header protocol field of 1):

- ICMP can carry error information. For example, if a source is trying to reach a destination that is not available, ICMP messages can be returned to the source indicating this fact.

- ICMP can be used to determine if a device is present. The well-known utility "ping" uses ICMP for this purpose.

- ICMP can be used to figure out what is the largest IP datagram size that can be supported by the network. This is useful to reduce the amount of fragmentation that is done.

We have been discussing IP Version 4, as this is the most widely deployed version of IP by a large margin. However, a new version (IP V6) is available and most commercial operating systems do support it. IP V6 addresses two main limitations with IP V4:

- It expands the address space from 32 bits (~4 billion addresses) to 128 bits (or a mind-staggering 340,282,366,920,938,463,463,374,607,431,768,211,456 addresses). This is in reaction to a perceived shortage of IP V4 addresses due to the exponential growth we discussed earlier. In fact, the use of the private address space in IP V4 with network address translation (NAT) has made this problem less important (although NAT introduces other problems).

- IP V6 introduces the concept of a flow ID for each datagram. A flow ID is akin to the IP V4 type of service but has a larger space (20 bits). It is used to allow for end-to-end quality of service.

2.5 The TCP/IP Transport Layer

The TCP/IP suite defines two transport layers: a connectionless and unreliable transport protocol known as the User Datagram Protocol (UDP) and a connection-based/reliable protocol (Transmission Control Protocol or TCP). Both of these run on top of the IP layer discussed previously. As is illustrated in Figure 2.2, these protocols map into roughly the OSI transport and network parts of the OSI stack.

Both these protocols also add another component to the IP address we discussed above. This component is known as the transport port, and if the IP address is likened to a street address, the port is the equivalent of an apartment number. Specifically, port numbers allow multiple IP connections between two IP addresses (i.e., two devices). This is needed, for example, when multiple applications running on a device need to communicate with their counterparts on the other device. Some port numbers are well-defined (meaning an application has registered the use of that port—for example, the Session Initiation Protocol used in Voice-over-IP call signaling has registered the port 5056 for its use). For other application protocols the port number may be randomly selected (as is the case for the Real-Time Protocol used to carry Voice-over-IP media).

2.5.1 Transmission Control Protocol (TCP)

For reliable, full-duplex, connection-based transport of data between two endpoints, the Transmission Control Protocol is the way to go. Unlike UDP, which takes a user-provided data chunk and passes it on to the IP layer, TCP is a stream-oriented protocol. To TCP, user data is a continuous stream of bytes (referred to as octets in TCP parlance). Actual data boundaries are determined entirely by the application and not necessarily visible to TCP. In particular, the contents of a TCP packet on the network will not always correspond to the application data boundaries.

TCP accepts a stream of bytes from the application and decides internally when it is a good time to package this data into a TCP packet and send it. The packet (also called a TCP segment) will contain a TCP header, the contents of which are defined in Table 2.2, and data.

Table 2.2: TCP Header Fields

Field	Size (bits)	Purpose
Source port number	16	TCP port from which this data is sent.
Destination port number	16	Destination TCP port for this data.
Sequence number	32	Sequence number of the 1st data octet in this packet.
Acknowledgment number	32	Sequence number of the last data octet that was received. Every TCP segment can contain an acknowledgment (the ACK bit flag indicates if this field is valid).
Header length	4	The number of 32-bit words in the TCP header, typically 5.
Reserved	6	
1-bit flags	6	• Urgent (rarely used) • ACK (indicates that the acknowledgment number in this header is valid) • PSH (set if sender wishes receiver to push this data to the application as quickly as possible) • RST (reset the connection) • SYN (synchronize sequence numbers—set at connection initialization) • FIN (set to indicate that the session should be ended)
Window size	16	This is used for end-to-end flow control and gives the number of maximum octets that the sender is prepared to receive.
TCP checksum	16	Checksum of the segment.
Urgent pointer	16	Offset to sequence number to find urgent data (valid when Urgent Flag is set). Rarely used.
{Options}		Various. The most common are: • No operation • Maximum segment size (MSS) • Windows scale factor • Timestamp

TCP is a reliable transport protocol. It accomplishes this by using a positive acknowledgment scheme (carried in the acknowledgment field of each TCP header). Unlike some protocols where each packet is acknowledged, TCP takes a more conservative approach. With TCP, each octet is given a unique 32-bit sequence number. In each segment, the sequence-number field gives the sequence number of the first octet in the segment. Using this and the length of the packet, the receiver can figure out the sequence number for any octet in the segment. The receiver then acknowledges the receipt of the segment by setting the Acknowledgment field (to the sequence number of the last octet it received plus one) of the next segment that it sends (and setting the ACK 1-bit flag). This ACK can be embedded in a data segment that the receiver is sending or, if the dataflow is currently unidirectional, as a standalone packet. Each side of the TCP connection picks a random starting transmit sequence number at the TCP session initialization (which we will discuss shortly). If more than 2^{32} octets are transmitted, the sequence number simply wraps.

To see how TCP provides reliability, consider the example in Figure 2.5. Device A sends data to TCP segments to B (starting with sequence number 123). In the example, the connection data flow is unidirectional from A to B, with each segment holding 100 octets. Device A is able to send multiple segments without waiting for an acknowledgment from Device B (the actual number of octets that it can have outstanding is determined by the window size in the TCP header carrying acknowledgments from Device B). Assume segment 3 is lost. Since A can send ahead, it will continue to send octets (assuming the window is not reached). Eventually an acknowledgment will be returned from Device B indicating that it can only acknowledge through sequence number 322 (B can save octets 423–522 internally but cannot forward them to the application until the gap is repaired). The reception of the acknowledgment from Device B tells Device A that it needs to retransmit the lost segment. However, Device A has various strategies on what to do after retransmitting the missing segment; for example, it can wait to see if Device B acknowledges all the other outstanding segments before sending any more, or it can continue sending from octet 523 (which is illustrated in the example). Part of this "flexibility" is due to the fact that the TCP does not provide a selective acknowledgment or negative acknowledgment mechanism.

Part of the retransmission strategy used by a TCP implementation is determined by the window-size field of the TCP header, W. As we mentioned above, the window size is used by the receiver to control the number of octets the transmitter can send before needing to see an acknowledgment. Essentially, if the receiver has acknowledged up to sequence number T and the window size given by the receiver is W, then the transmitter can only send up to octet $T+W$ before having to wait for the next acknowledgment from the receiver. The window size is also used for flow control. Various algorithms have been defined to optimize how the sender and receivers should use windows. One important algorithm is known as slow start, in which the sender begins by sending just one segment and waiting for an acknowledgment before

Device A   Device B

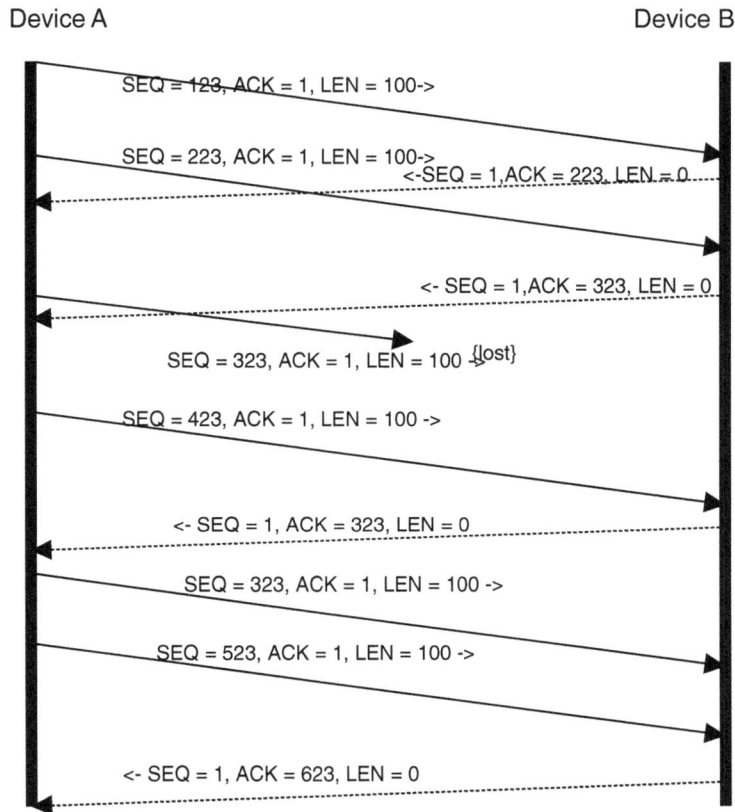

SEQ = 123, ACK = 1, LEN = 100->

SEQ = 223, ACK = 1, LEN = 100->
<-SEQ = 1, ACK = 223, LEN = 0

<- SEQ = 1, ACK = 323, LEN = 0

SEQ = 323, ACK = 1, LEN = 100 {lost}

SEQ = 423, ACK = 1, LEN = 100 ->

<- SEQ = 1, ACK = 323, LEN = 0

SEQ = 323, ACK = 1, LEN = 100 ->

SEQ = 523, ACK = 1, LEN = 100 ->

<- SEQ = 1, ACK = 623, LEN = 0

Figure 2.5: Example of TCP Retransmission

sending any more (even though the receivers window may be larger). After each successful acknowledgment, the sender doubles the number of segments it sends before waiting for an acknowledgment (up to the receiver's advertised window). This lets the sender "test the waters" as to what the optimal window should be.

TCP is a connection-oriented protocol. What this means is that a connection initialization sequence is required before any TCP data can be sent between two devices. This startup sequence is known as the TCP three-way handshake, and is illustrated in Figure 2.6.

The TCP connection setup is asymmetrical. The passive (or server, Device B) side of the connection begins by listening for connection requests on a particular TCP port. For many applications, the port is well known (e.g., TCP 5056 for the Session Initiation Protocol Application). The active side of the connection (client, Device A), sends an initial TCP segment to the destination IP address and port. The server then responds with a segment. Finally, the client acknowledges the server, thus completing the three-way handshake. The segments

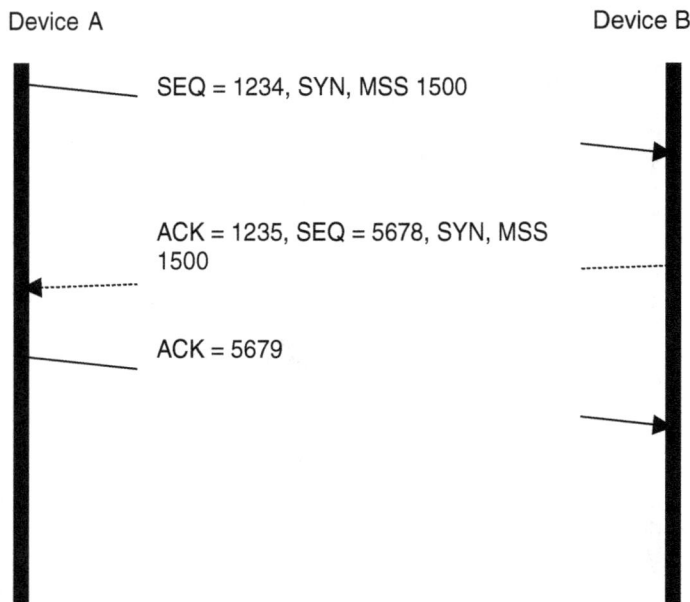

Figure 2.6: TCP Three-way Handshake

exchanged in this initialization are special in that they contain some information in the TCP header, not present in other (data) segments:

- SYN bit is set in the flags header field. This indicates that the segment is a TCP connection request or response.

- The sequence numbers are set to some [random] value. These are the initial sequence numbers that the client and server will use to number the octets that they send.

- The MSS option can be present to indicate the maximum segment that the client or server is prepared to receive.

These segments may or may not contain real data. The SYN flag takes up one sequence number, hence the need for the acknowledgment.

Session teardown in TCP is the most complicated aspect of the protocol, requiring four segments to complete the transaction (as illustrated in Figure 2.7).

Essentially, the termination sequence consists of two "half-close" operations. Each side of the connection must tell the other that it has finished transmitting data by sending a segment with the FIN flag set. The other side must acknowledge this segment for the sending side to officially close its side. It then must wait for the other side to close its half, via the same

Device A   Device B

SEQ = 4321, FIN, ACK = 9876

ACK = 4322

SEQ = 9876, FIN

ACK = 9877

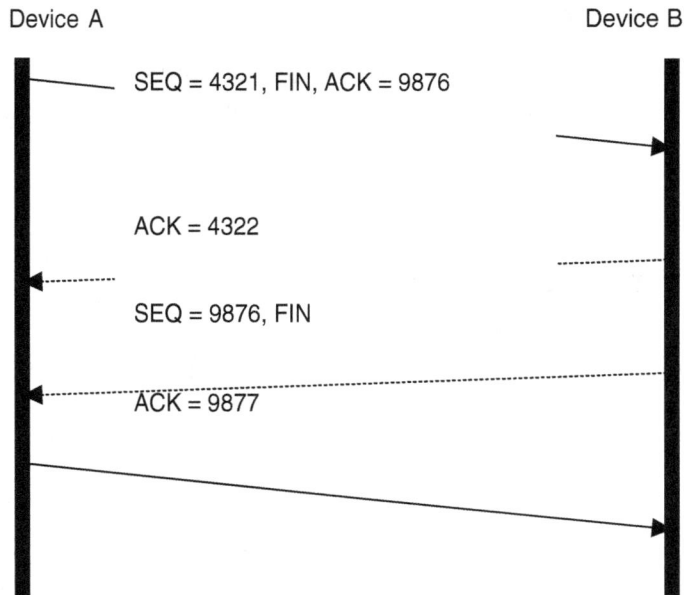

Figure 2.7: TCP Session Termination Sequence

handshake. Note that Device B could still send data (which Device A would need to acknowledge) before it sent its FIN segment. TCP also defines various timers to handle the case where
the remote device fails to respond to the initialization or termination sequences. In the case of
session termination, TCP requires that each side, after the receipt of the other side's FIN, stay
alive for a period of time known as the TIME_WAIT state (or 2MSL wait state). The purpose
of this wait is to ensure that any stray packets from the connection that has just been terminated don't arrive. The duration of this wait state is defined to be two times the maximum
segment lifetime (MSL—hence, the name 2MSL wait state). A typical value for this wait
period is 2 minutes.

To conclude the discussion of TCP, we can take note of its advantages and disadvantages.
Under the "plus" column we can state that TCP provides a reliable transport layer with some
built-in extras including flow control. On the negative side, an application using TCP must
forgo a certain amount of control on how its data is packaged into IP datagrams, and the exact
timing of when the datagrams get sent and ultimately delivered to the end application. As we
will discuss in the next chapter, applications with real-time requirements will typically use
UDP instead.

2.5.2 User Datagram Protocol (UDP)

UDP is a lightweight protocol and in reality provides very little on top of the IP layer. It takes a chunk of user data and adds a UDP protocol header (of 8 bytes) before passing it to the IP layer. The UDP header consists of the fields in Table 2.3:

Table 2.3: UDP Protocol Header

USP Header Field	Size (bits)	Purpose
Source port	16	This gives the port used by the sender to transmit the datagram.
Destination port	16	This gives the destination port for the datagram.
Length	16	This gives the length of the UDP datagram.
Checksum	16	This gives a checksum over the user's data, UDP header and parts of the IP header. This can be used to verify that the entire datagram has not been corrupted during transmission through the network.

Often the UDP checksum is ignored and hence the most important fields are the port numbers which, as explained above, allow multiple connections between two IP addresses. Given that UDP is pretty lightweight and does not provide many of the functionalities of TCP, why would anyone use it? The answer is that, for certain applications, we are concerned with getting the data to the destination as quickly as possible and are willing to tolerate a certain amount of unreliability. Also, these applications need the second level of addressing provided by the UDP port. As we shall see, UDP is used to transport Voice-over-IP media for this very reason.

2.6 Other TCP/IP-Based Protocols

In this section we will present a survey of other protocols that utilize the TCP/UDP/IP layers discussed previously. These include both IP companion protocols such as RIP as well as application layer protocols such as Telnet. Table 2.4 gives a summary of a subset of the TCP/IP protocol family.

Table 2.4: TCP/IP-based Protocols

Protocol Name	Based On	Purpose	Data Format	Reference
ICMP	IP	IP Control Messages	Binary format	RFC 792
IGMP	IP	Multicast Group Management	Binary format	RFC 1112
RIP, RIP v2	UDP/IP	Distribute Routing Information	Binary format	RFC 1058 (v1), RFC 2453 (v2)
OSPF	IP	Open Shortest Path First Routing Protocol	Binary format	RFC 2328 (v2)
BGP	TCP	Border Gateway Protocol		RFC 1771 (V4)
DHCP	UDP	Dynamic Host Configuration Protocol—used to provision IP address and related information to remote devices	Binary format	RFC 2132
Telnet	TCP/IP	Remote Terminal Access	ASCII text	RFC 137
Rlogin	TCP/IP	Remote Terminal Access	ASCII text	RFC 1282
DNS	TCP/IP	Domain Name System—provides mapping between IP addresses and hostnames	Binary format	RFC 1034
SMTP	TCP/IP	Mail Transfer	ASCII	RFC 1213
TFTP	UDP/IP	Simple File Transfer	Binary control format, arbitrary data format	RFC 1350
FTP	TCP/IP	File Transfer	ASCII control format, arbitrary data format	RFC 959
NFS	UDP/IP	File Transfer (network file system)	Binary format, using external data representation format (RFC 1014)	RFC 1094
SNMP	TCP/IP	Remote Device Management	ASN.1 encoding	RFC 1905 (v2)
HTTP	TCP/IP	HyperText Transfer Protocol	ASCII for control	RFC 2068
SNTP	UDP/IP	Simple Network Time Protocol	Binary	RFC 2030 (v4)
Session Initiation Protocol	TCP/IP or UDP/IP	VoIP Call Signaling	Text based	RFC 3261

Table 2.4: TCP/IP-based Protocols (continued)

Protocol Name	Based On	Purpose	Data Format	Reference
Media Gateway Control Protocol	UDP/IP	VoIP Call Signaling	Text based	RFC 2705
H323	TCP/IP	VoIP Call Signaling	ASN.1 encoded	ITU Recommendations H series
RTP	UDP (primarily)	Real-Time Transport Protocol Media Transport	Media dependent	RFC 3550

TCP/IP-based protocols can be broadly characterized as protocols that work "under-the-hood" to provide the base TCP/IP services and pure application-level protocols. In the former category, we have the companion protocol to IP (ICMP) that we discussed above, and the various routing-related protocols (RIP, OSPF, BGP, IGMP etc.). We can also lump the Domain Name Look-up protocol (DNS) and the Dynamic Host Configuration Protocol (DHCP) in this category.

DNS is an important protocol in any IP application because it provides a means to map (or re-solve) a service or a device name (referred to as a fully qualified domain name or FDQN) to an actual IP address and vice-versa. DNS is a distributed database; each site (e.g., company, cam-pus) maintains information on its IP hosts. The DNS protocol defines the mechanisms used to query these databases. Typically, any IP service will be preceded by a DNS look-up operation.

DHCP is an even more fundamental protocol. In any large IP network, it is infeasible for each IP device to be assigned an IP address statically. Instead, the DHCP protocol allows an IP device (DHCP Client) to request an IP address from the network (using the IP protocol broadcast capability). DHCP Servers respond to the request with a free IP address. The client then selects an IP address from one of the responses and indicates its choice to that server. The server then replies with a final acknowledgment and marks that IP address as "in-use." IP addresses obtained from DHCP typically have a lease time associated with them. Clients are responsible for re-requesting the IP address before the lease expires. In addition to IP addresses, DHCP responses will contain other important information such as the *netmask*, default gateway address, the address of the DNS server to use, and possibly the address of other servers (e.g., provisioning, time).

File transfer and terminal access are two of the earliest pure application protocols. TFTP is based on UDP and uses a binary message format. TFTP is a client/server protocol and defines five message types for the client and server to use: read file request, write file request, [file] data, acknowledgment, and error. As UDP is an unreliable transport protocol, TFTP has to handle lost or duplicated packets itself through the use of timers and the acknowledgment and error message types. TFTP is commonly in IP devices (such as VoIP terminals) to retrieve provisioning data from a common server.

FTP, on the other hand, is based on TCP and thus runs over a reliable transport layer. FTP has several characteristics that are common to many IP application protocols, and so are worth mentioning. The first characteristic is that it is a request/response protocol with control messages (e.g., messages that are sent by the client and server to set up, start, stop, etc.) defined in the protocol, all are passed in ASCII (i.e., human-readable, text-character) format. There are two advantages for this. First, being human readable, it is easy to understand and debug the protocol operation. Secondly, as ASCII is a standard representation, it makes for a simple presentation layer. Using text messages to pass control information means that a machine-independent data-representation format does not need to be defined. This keeps the implementation simpler.

A second characteristic of FTP is the use of [ASCII] numeric codes in response messages. The numbering convention used by FTP is followed with other applications as well. For example, responses in the 100 to 200 range indicate a preliminary reply. These give intermediate status of the client command but do not indicate that the command has completed. Responses in the 200–300 range indicate that the command has completed successfully and a new transaction can be started. Responses in the 300–400 range also indicate success, but that the server expects another command related to the transaction to be sent next. Failures are indicated by responses in the 400–500 or 500–600 range, the difference being that 400–500 indicates a transient failure (e.g., server is temporarily busy) while 500–600 indicates a permanent failure such as "file not found."

A final characteristic is that the transport port used for control messages and the actual [file] data is different. FTP has a "port" command, where the client indicates which TCP port it expects the file to be sent to.

Mail transfer, HTTP, and several of the VoIP protocols all share many of these characteristics.

2.7 Conclusion

In this chapter we have seen an overview of packet data networks based on the Internet protocol suite (TCP/IP). With this background, we can next look at Voice over IP (VoIP) applications in detail.

References

[1] *TCP/IP Illustrated*, Volume 1, The Protocols, W. Richard Stevens, Addison-Wesley Professional Computing Series, 1994

[2] *Computer Networks*, A Tanenbaum, Prentice-Hall, Inc., 1981.

[3] *TCP/IP Illustrated*, Volume 2, Gary R. Wright and W. Richard Stevens, Addision-Wesley Professional Computing Series, 1995

Voice over IP

3.1 Introduction

The previous chapters have defined the PSTN and Internet in some detail. As we saw, these two physically and logically separate networks were designed with very different design goals in mind and for quite some time they have continued to exist separately in real-world deployments, as shown in Figure 3.1.

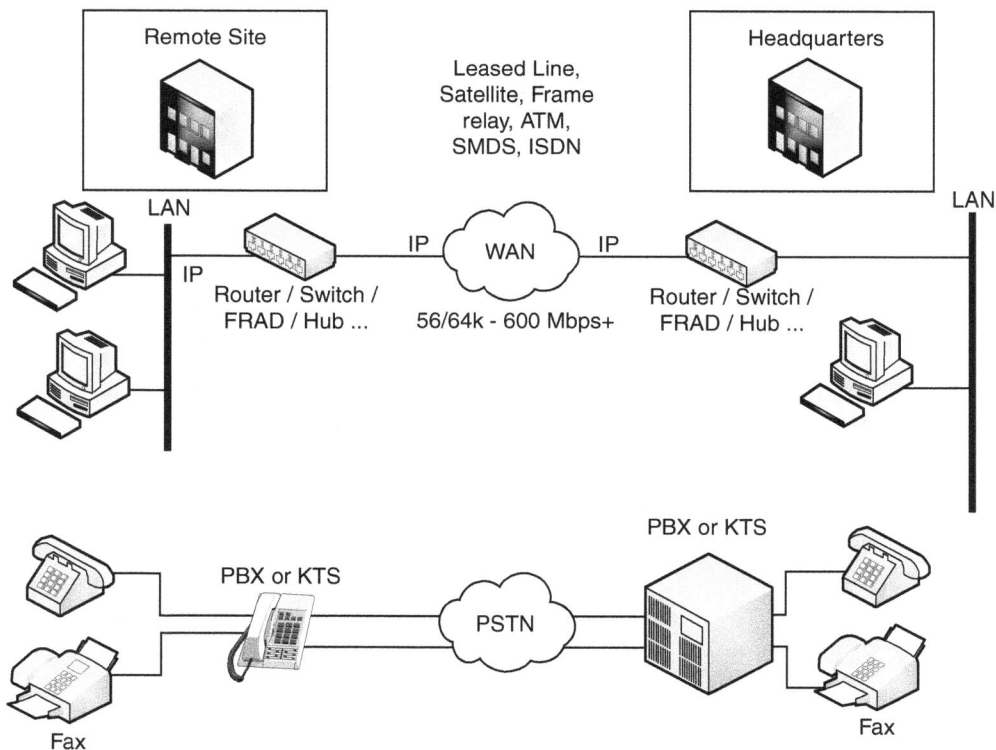

Figure 3.1: Traditional Architecture: Separate Networks for Voice and Data

Probably the earliest convergence of the PSTN and Internet was dial-up connections where end users used a modem and their phone line to connect to the Internet. However, we are concerned here with a different model of convergence, in which the Internet is used for carrying voice calls. Voice-over-IP (VoIP) can be briefly defined as the technology that allows the use of the IP protocol(s) to carry voice signaling and media traffic. We look at how VoIP works in this chapter, but first we define the need for VoIP.

3.1.1 Motivation for VoIP

As anyone who has worked in the industry long enough would tell you, if two distinct networks exist, somebody somewhere will find a reason (and implement a protocol/product) to connect the two networks. In the case of PSTN and Internet, this convergence came in the form of VoIP. The primary motivation for VoIP was cost savings in the near term. By using VoIP, the end users could bypass toll switches and save on call costs. This was especially relevant for international or long-distance calls and hence one of the earliest adoptions of VoIP was in international/long-distance "calling cards."

Soon, the long-distance carriers also realized the potential of this cost savings and adopted VoIP to step to a single converged network. This also meant cost savings for the carriers, which were passed on to the end user in the form of falling long-distance/international toll charges.

It is interesting to note that at least some part of VoIP cost savings stem from the fact that VoIP service providers are usually not regulated in most countries and do not have to pay a licensing/operating fee to the government. In fact, some governments have prohibited the use of VoIP in their countries to ensure that the licensing/operating fee paid by the telephone service providers can be justified. It is possible that in the near future VoIP will become a regulated industry. However, most governments have not yet implemented a mechanism to charge an operating fee for VoIP providers. If and when this happens, some of the cost-saving advantages of VoIP will be compromised.

However, other cost-saving features of VoIP would continue. For example, by using low-bit-rate codecs, voice can be compressed to use much less bandwidth (typically < 20 kbps) than that used by PSTN (64 kbps). By virtue of operating in a connection-less network, bandwidth usage can be optimized so that no bandwidth needs to be reserved as in connection-oriented networks such as PSTN. Lower costs also result from maintaining a single network instead of two physically and logically separate networks.

Besides lower costs, additional revenue from new services and applications arising from integrated data and voice would continue to be major motivators for VoIP adoption. For example, by using advanced encryption, VoIP can provide secure voice paths to paying subscribers.

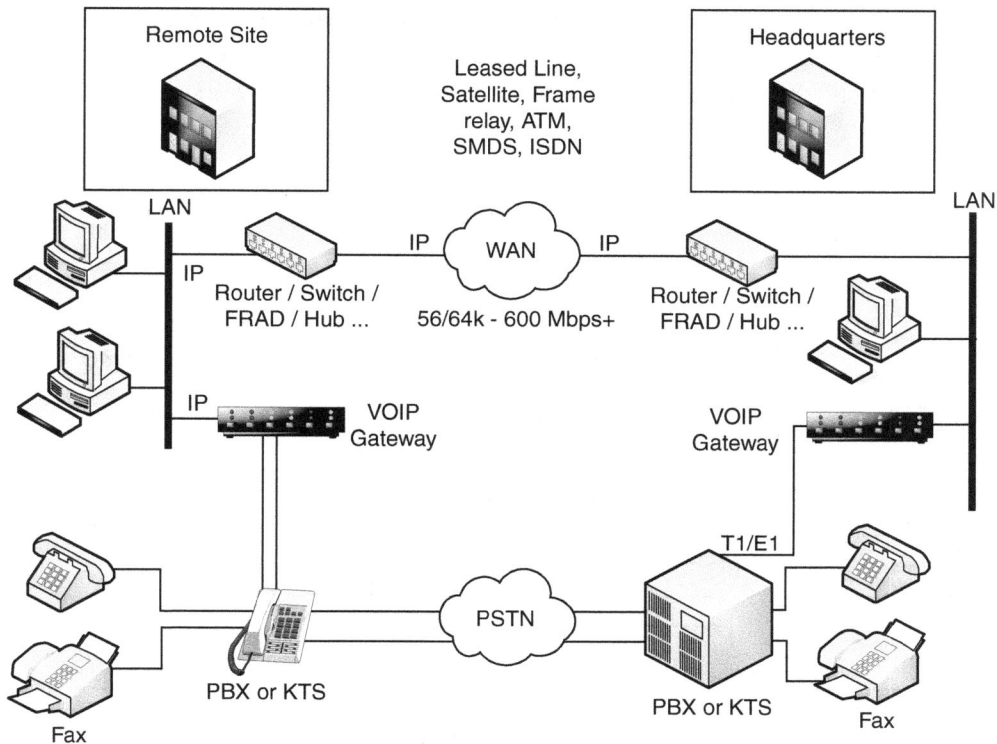

Figure 3.2: Convergence of Telephony and Data

3.1.2 Challenges in VoIP

Now that we know the motivation for using VoIP, we can get down to examining the challenges that arise in using IP networks for voice communication. To understand these challenges, it is important to keep in mind that IP-based networks like Internet are connectionless, packet-switched networks, whereas voice has traditionally been carried over connection-oriented, circuit-switched networks that were dedicated to carrying voice communication.

3.1.2.1 Delay (Latency)

One of the most important challenges of VoIP is delay. Delay can come in two flavors: absolute delay and variable delay (aka jitter). Due to the real-time nature of voice communication, an absolute delay of more than 200–250 ms in the voice path can make the voice quality unacceptable. Hence, for VoIP to deliver acceptable voice quality, designers and service providers must meet this requirement. As shown in Figure 3.3, there are various sources of delay in a VoIP network. On the transmit side endpoint, the packetization time, the encoding time and transmission time constitute the bulk of this delay. In the network, besides the propagation delay, routers introduce processing delay, queuing delay and transmission delay—sometimes together known as the switching delay. Finally, at the receiver endpoint, besides the processing delay, a play-out buffer also adds to the delay. To understand the concept of play-out buffer, we need to understand the other aspect of network delays, which is variable delay.

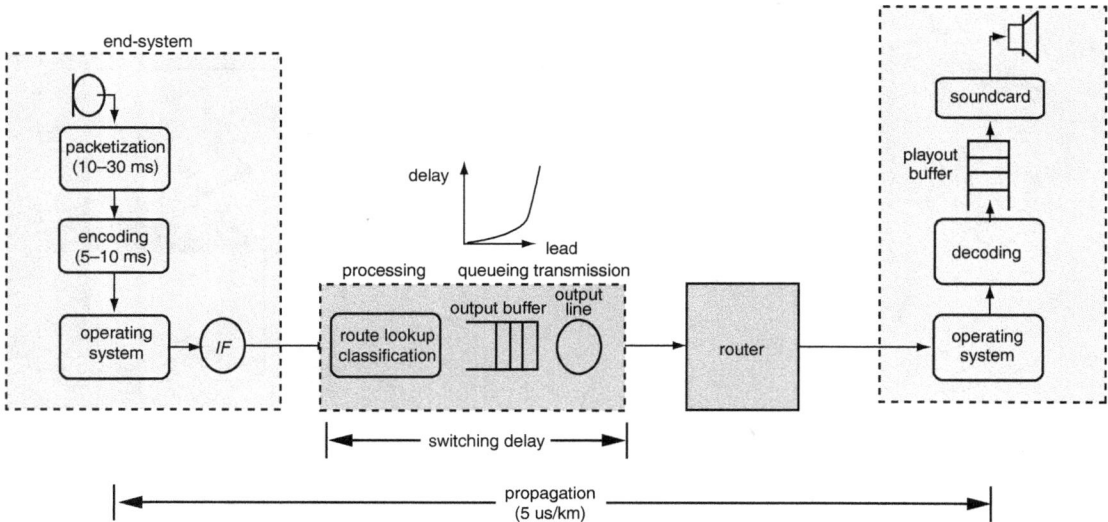

Figure 3.3: Delay in IP Networks

Refer to the switching delay shown in Figure 3.3. As shown, the queuing delay in routers depends on the load the routers are handling. Since this load varies depending on a host of factors, such as time of day, the end-to-end delay experienced by packets is variable too. For real-time applications like voice, this is a big problem. Since the receiving endpoint is playing out voice packets in real time (as it receives them), what should it do when an expected packet is delayed? Doing nothing would mean a perceived disturbance for the receiver. To work around this problem, the receiving endpoint does not play out each packet as it is received; instead it delays each packet by storing it in a play-out buffer. Selecting the extent of this delay (aka depth of the play-out buffer) is an engineering trade-off—larger values mean more protection against jitter (variable delay) but at the cost of increased absolute delay. In any case, jitter is a big problem for VoIP.

3.1.2.2 Packet Loss

It is often said that delay is more important than packet loss in VoIP, or that packet loss is much more acceptable in VoIP than in data applications. However, such statements should be very carefully qualified. It is true that voice communication can tolerate some (typically < 5%) packet loss due to the inherent redundancy in voice. However, the tolerance limit varies from one VoIP implementation to another depending on what codecs are being used and whether or not some packet loss concealment (PLC) schemes have been implemented. Any packet loss beyond this threshold leads to significant deterioration of voice quality and, hence, packet loss should be minimized in VoIP networks.

3.1.2.3 Available Bandwidth

Bandwidth is one of the most important and one of the most confusing topics in telecommunications today. If you stay current with the telecommunications news, you would have come across conflicting reports regarding bandwidth. There are a lot of people claiming "bandwidth is cheap" and probably as many people claiming "it is extremely important to conserve bandwidth." So, what's the deal? Do networks today have enough bandwidth or not? The problem is there is no single correct answer to that. The answer depends on where you are in the network. Consider the core of the IP and the PSTN networks: the two most widely deployed networks today. The bandwidth available at the core of these networks is huge, thanks to fiber optics; bandwidth, therefore, is cheap at the core of the network. Similarly, the dawn of 100-Mbps and gigabit Ethernet has made bandwidth cheap even in the access network (the part of the network that connects the end user to the core). The wireless medium, however, is a little different and follows a simple rule: bandwidth is always expensive. This stems from the fact that in almost all countries the wireless spectrum is controlled by the government. Only certain bands of this spectrum are allowed for commercial use, thus making bandwidth costly in the wireless world. All protocols designed for the wireless medium therefore revolve around this central constraint. Therefore, VoIP over wireless (i.e., VoWLAN) must also try to minimize bandwidth consumption, which depends on codecs (see section 3.2) and packetization periods (see Chapter 5).

3.2 Putting Voice Over Internet

The PSTN can be logically separated into a signaling subsystem and a media-transport subsystem. Using Internet to carry voice calls also requires implementing signaling and media transport. We look at signaling in section 3.4 and media transport in section 3.5. This section provides a very high-level view of how voice traffic can be transported over Internet.

Sampling: User speech is converted (by collecting voice samples) from analog voltages from the microphone into digital samples. Since human speech's range is less than 4 kHz, sampling is typically done at 8 kHz.[1] However, 16-KHz sampling is also becoming popular for "CD quality" sound.[2] With 8-bit sampling at 8 kHz, 8000 8-bit samples are generated each second, thus leading to a 64-kbps voice channel. With 16-bit sampling, 8000 16-bit samples are generated, leading to a 128-kbps voice channel.

---

[1] Nyquist's theorem: An analog signal waveform may be uniquely reconstructed, without error, from samples taken at equal time intervals if the sampling rate is equal to, or greater than, twice the highest frequency component in the analog signal.

[2] There are three common digitization standards today: 8-bit mu-law, 8-bit A-law and 16-bit linear.

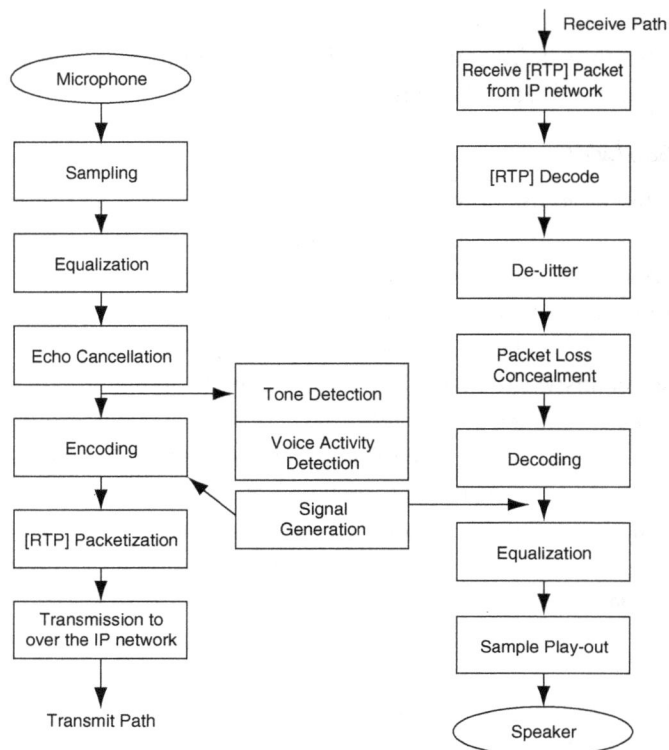

Figure 3.4: Steps Involved in VoIP

Equalization: The purpose of this step is to dynamically adjust the samples to counter known limitations in the microphone. This is a voice-quality enhancement feature; simpler and cheaper implementations of VoIP may not implement this step.

Echo Cancellation: The purpose of echo cancellation is to remove residuals of the voice received from the far-side or remote-end (and being played out to the user) from being accidentally transmitted back to the sender. Echo can occur for several reasons. In the case where the microphone is part of a traditional PSTN black phone, it can occur as a result of the electrical interface between the handset and the PSTN tip/ring wiring. This is known as the hybrid, and the cancellation technique used is referred to as *line echo cancellation*. In the case that we are most interested in, namely a voice over WLAN handset, echo will occur as a result of acoustical coupling between the handset speaker and microphone. The techniques used in this case are referred to as *acoustic echo cancellation*. Echo cancellation is present in non-VoIP telephony systems also; the problem is more complicated with VoIP because of the longer network delays. With longer delays echo becomes more of a degradation. Almost every VoIP implementation employs some sort of echo-cancellation mechanism, but echo cancellation continues to be one of the primary areas of product differentiation among various VoIP implementations.

Encoding: These digitized samples are typically collected in groups, known as voice frames. A voice frame can contain anywhere from 1 ms worth of samples to 30 ms worth of samples, depending on the codec being used. The samples are now ready for encoding. The processing in this step is defined by the codec used for the call. A codec defines the mathematical rules for translating the raw samples into a compressed data stream. A partial list of popular audio codecs is given in Table 3.1:

Table 3.1: Common VoIP Audio Codecs

CODEC	Bit Rate	Bits per voice packet
G711	64 kbps	640 bits for 10 ms
G726-32	32 kbps	320 bits for 10 ms
G729ab	8 kbps	80 bits for 10 ms
G729e	12 kbps	120 bits for 10 ms
G728	16 kbs	160 bits for 10 ms

As is obvious from this table, this step leads to one of the primary advantages of VoIP: bandwidth savings. Consider, for example, G729ab—using this codec, we can compress the voice stream from 64 kbps down to 8 kbps.

Digit Relay: As we discussed in Chapter 1, the PSTN can be logically separated into a signaling subsystem and a media-transport subsystem. The signaling subsystem is CCS (common channel signaling) in the core network and CAS (channel associated signaling) in the access network. VoIP implementations therefore need to implement these signaling subsystems too. In the access network, the trivial solution is to do nothing—i.e., treat DTMF tones as speech and encode them just as speech. However, this approach fails when low-bit-rate codecs are used, since these codecs have been optimized for human speech. Therefore, DTMF (aka digit) relay is often used to solve this problem. As explained in RFC 2833, this involves detecting that a digit has been pressed and then sending this information out-of-band from the voice path in a separate type of packet.

Voice Activity Detection (VAD): The purpose of VAD is to detect when active speech is present in the signal. The underlying concept is based on the observation that human speech consists of periods of silence, when the person on the other end is talking or between sentences and words. During these periods of silence, therefore, it is possible to stop voice-packet transmission, thus saving bandwidth. Often VAD is embedded as part of the coding block (codec) but it may also be implemented as a separate unit or block.

Redundancy: As explained in section 3.1.2, one of the major challenges for VoIP is to cope with packet loss, which is much more common in IP networks than it is in connection-oriented networks like PSTN. To proactively protect against packet loss, which is widespread in IP-based networks, VoIP implementations sometimes use redundancy in the voice path. This redundancy is achieved by each packet carrying information content of *N* previous packets as well. Now, if any one of these *N* packets gets lost, the current packet can be used to recover those packets at the receiver. The higher the value of *N*, the higher the bandwidth used—thus, redundancy provides protection against packet loss at the cost of higher bandwidth consumption.

Packetization: The voice frames from the codec process are now ready to be grouped into [RTP] packets. Multiple frames may be combined to form a single transmission packet. It is this packet which is then transmitted to the destination over the IP network. A key parameter in VoIP is the packetization period. Longer packetization periods are more efficient but introduce more delay. A packetization period of 20 ms is common.

On the receive side, the reverse processes are carried out. The VoIP terminating unit is responsible for extracting the encoded samples from the received voice [RTP] packets, decoding them, performing dejitter on them, and playing them out to the user speaker/handset/headphone. Additionally, on the receive side it is also important to account for tone detection.

Tone Detection: The samples extracted from the RTP packets may be analyzed by one or more detectors. Tone detectors look for particular combination of frequencies in the transmitted signal. An example is a dual-tone multifrequency (DTMF) detector, which looks for the dual-tone signal generated when a digit is dialled from a PSTN or black phone. Other tones that are commonly looked for in VoIP are fax and modem signaling tones. These are necessary so that the VoIP terminal can determine that a FAX machine or modem is involved in the call and make adjustments to ensure that the fax or modem call goes through successfully.

3.3 VoIP Architectures

VoIP deployments can have various architectures. The simplest way to classify these architectures is to understand where voice transitions from the PSTN to the Internet.

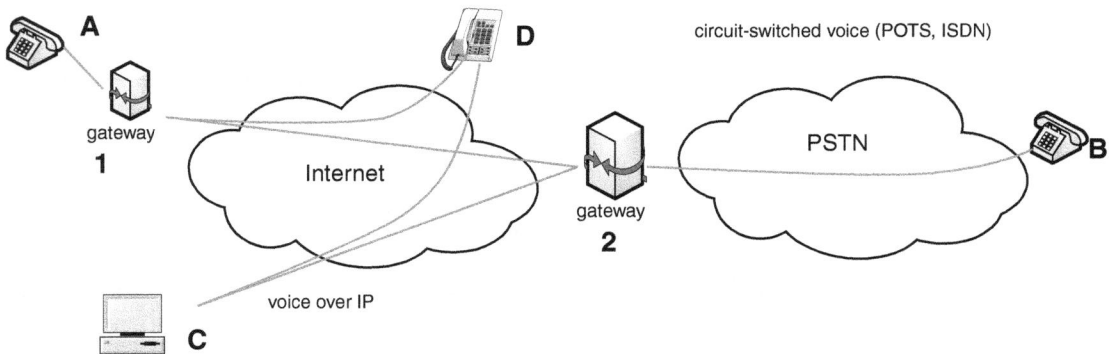

Figure 3.5: Telephony—PSTN and VoIP

Figure 3.5 shows a very high-level classification of VoIP architectures. It consists of four endpoints of communication, two of which are PSTN devices (A and B) and the other two are Internet devices (C and D). Note that a PSTN device, most often, refers to a plain old telephone (aka black phone) whereas an Internet device may be software running on a PC (aka a soft IP phone) or a physically separate device like a VoIP handset. Now, let's consider some scenarios:

C-D voice call:

In this scenario, both endpoints are Internet devices and hence the voice call never leaves the IP domain. This makes things a little simpler. Signaling protocols (see section 3.4) typically used in such scenarios are peer-to-peer, for instance, SIP or H.323.

A-D voice call:

In this scenario, we have one endpoint as an internet device (D) and one endpoint as a PSTN device (A). To enable communication between a VoIP device and a PSTN device, a VoIP gateway is used.

Here, it is useful to understand the concept of the term "gateway" since it is used extensively in VoIP. A gateway is a logical entity that interconnects two heterogeneous networks, such as PSTN and IP in the case of VoIP. Gateways can be classified based on different criteria. Based on functionality,[3] a VoIP gateway can be a signaling gateway (aka softswitch) or the media gateway. The former is responsible for ensuring interworking between VoIP signaling (see section 3.4) and PSTN signaling (see Chapter 1). The signaling gateway does this by translating PSTN signaling commands to corresponding messages in the VoIP signaling domain. It is therefore the responsibility of the signaling gateway to ensure that a voice call gets established between A and D. On the other hand, the media gateway is responsible for ensuring that the voice path's transition from the PSTN domain (A-1) to the IP domain (1-D) is transparent to A and D. It does so by reformatting the packet headers and acting as a translator (different codecs), among other things.

Another classification of gateways is based on the size/capacity of the gateway. A residential gateway is one deployed at the customer premise. Such residential gateways connect on one side to the RJ45/home telephony wiring infrastructure and on the other side to the internet. These devices allow the user to plug in a conventional PSTN phone and make calls over VoIP instead of the PSTN. An example is Gateway-1 in Figure 3.5. Next in the hierarchy are various forms of small to medium capacity gateways deployed in office complexes. Such gateways can enable different numbers of conventional PSTN phones to make VoIP

---

[3] Note that this is a logical division. A single physical device may serve both as the signaling gateway and the media gateway.

calls. Finally, there are trunking gateways, which are very high-capacity gateways, typically deployed by telephony service providers to offload calls from the PSTN to the IP network. In Figure 3.5, Gateway-1 may be a residential gateway sitting at A's customer premises.

A-B voice call:

In this scenario, we have two PSTN endpoints involved in a voice call. Traditionally, such a voice call would have travelled completely over the PSTN network. However, for reasons discussed in section 3.1.1, the telephony service provider in Figure 3.5 has decided to deploy a trunking Gateway-2 to offload the voice call to the Internet. The voice call travels as VoIP between Gateways 1 and 2 and as circuit-switched telephony between A and 1 and B and 2.

Note that in Figure 3.5 we assume that the Gateways-1 and -2 act as both signaling and media gateways. This may not be the case in real-life deployments where physically separate entities are deployed for signaling and media gateways.

We now have an idea of how various VoIP entities (IP phones, residential gateway, trunking gateway, signaling gateway, etc.) are used to achieve different VoIP architectures. These VoIP architectures are differentiated primarily by where, if anywhere, voice transitions from the PSTN network from/to the IP network.

3.4 Signaling Protocols

Chapter 1 described the PSTN, which consisted of a signaling network (SS7) and a media-transport network (Class-5/4 Switches and Voice Trunks). The PSTN is a connection-oriented circuit-switched network where signaling is used to reserve dedicated resources between the communicating endpoints at the start of a voice call. These resources are used to carry voice traffic during the voice call and are released only when the voice call ends. The Internet, on the other hand, is a connectionless packet-switched network where each packet is routed independently.

The steps that need to be performed (in VoIP or in the PSTN) to set up a call, to manage the call, and to tear down the call are referred to as call signaling. The actual mechanism used to perform these actions is the call-signaling protocol. In this section we will discuss some of the more common call-signaling protocols used today in VoIP, concluding with the current industry leader, SIP. We will concentrate on protocols based on open standards (ITU, IETF), but it is important to recognize that there are also proprietary signaling protocols in wide-spread use. Skype and Cisco Skinny protocol are examples of these.

In general, all these VoIP call-signaling protocols are application-level protocols that sit on top of a transport layer such as TCP or UDP. In addition to the classification of open standards versus proprietary, the VoIP signaling protocols can be divided into central-ized protocols and distributed protocols. MGCP and Megaco are examples of centralized protocols. With centralized protocols, call state is maintained in centralized locations in the IP networks, known as call agents, soft switches or call-management servers (CMS). Endpoints (also known as remote gateways, access gateways, residential gateways, CPE devices, or media terminal adaptors) are responsible for following commands of the call agent and reporting events back to the call agent.

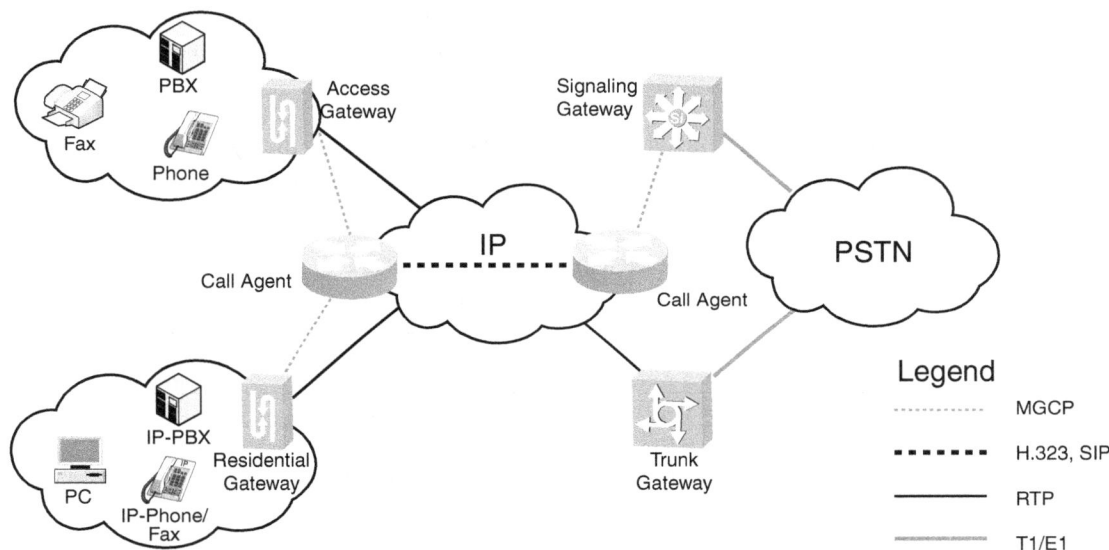

Figure 3.6: VoIP—High Level Architecture

Distributed protocols, as the name suggests, allow for call-state knowledge to be shared over multiple entities. A distributed protocol, such as SIP for example, can involve just two entities: the calling party and the called party. Each would have full knowledge of the call state. No centralized entity is required.[4] This is the case in Figure 3.6 where the two call agents use SIP/H.323 to communicate. As can be seen, distributed (SIP/H.323) and centralized (MGCP/Megaco) can exist together in a network.

3.4.1 Media Gateway Control Protocol

The Media Gateway Control Protocol (MGCP) is defined in RFC 3435. An important variant is specified by packet cable, and is referred to as MGCP network controlled signaling (or NCS). MGCP is a centralized protocol and consists of two components: the remote gateway (RG) and the call-management server (CMS). MGCP defines a remote procedure-like protocol to allow a CMS to control an RG.

---

[4] Note also that, in practice, distributed protocols such as SIP and H323 also involve centralized entities, each maintaining some degree of call state as well.

An RG may handle one or more media-terminating devices (or endpoints). MGCP uses a hierarchical e-mail-like naming convention for endpoints. For example, *aaln/1@my.mta.com* could refer to the first analog endpoint at the device my.mta.com. Wildcards are permitted, so the CMS could address all endpoints handled by my.device.com by using *aaln/*@my.mta.com*. Endpoints are one of the two main control objects in MGCP. For a given endpoint, the CMS will tell the RG what events to look for and what signals to apply. The second control object in MGCP is the call leg. A call leg, in the context of a normal call, can be viewed as a logical connection between the endpoint and the other end of the conversation. We will discuss events, signals and call legs a little more below.

MGCP call-signaling information is passed in MGCP messages. The main MGCP message types are defined in the table below. The majority of messages are from the CMS to the RG.

Table 3.2: MGCP Signaling Messages

Msg Type	Direction	Purpose
RQNT	CMS → RG	CMS uses this to tell RG what signals to play, and what events to report.
NTFY	RG → CMS	RG uses this to report events to the CMS.
AUDIT	CMS → RG	CMS uses this to determine the RG capabilities and status.
CRCX	CMS → RG	CMS uses this to create a media connection. A CRCX may contain an embedded RQNT.
MDCX	CMS → RG	CMS uses this modify a media connection. A MDCX may include an embedded RQNT.
DLCX	CMS → RG, RG → CMS	CMS uses this to delete a media connection. RG can also use this to indicate that he has auto-deleted the connection. A CMS issued DLCX may include an embedded RQNT.
RSIP	RG → CMS	Used by RG to register with the CMS.

MGCP messages are ASCII text and are sent over UDP transport protocol. Each message is given a unique transaction number and is responded to by an acknowledgment message containing the transaction number, a response code, and possibly other information. For example, a successful response to an RQNT message is indicated by the CMS receiving a 200 OK. MGCP defines a retransmission protocol to handle missing messages. Each RG, for example, is required to keep a message history list so that it can re-reply if it sees a duplicate transaction.

An MGCP message consists of a header containing the message type, a transaction number, the endpoint name, and the MGCP version. Each subsequent line contains a line identifier, followed by a colon, followed by line parameters. The list of important MGCP lines is given in Table 3.3.

Table 3.3: Key MGCP Message Lines

Msg. Line	Purpose
N:	Gives the address of the CMS where events should be reported.
X:	Request Identified, used to correlate reported events to requested events.
R:	List of events to look for and how to deal with them.
S:	List of signals (tones, etc.) to apply.
L:	Local connection options—this gives the list of codecs, packetization periods and other codec parameters that the CMS is recommending to use for the call.
C:	Call identifier—text string used to identify a call.
I:	Call leg identifier—text string used to identify a call leg. In MGCP a call will consist of one or more call legs.
O:	List of events to report to the CMS.
D:	Digit map that the CMS desires the RG to use.

3.4.1.1 MGCP Call Setup

The best way to understand MGCP is to see it in action. Figure 3.7 gives the call flow for a basic call from one RG to another.

Let us assume an RG (my.mta.com) containing one endpoint (aaln/1) is powered up and is initiating the call. After seeing an RSIP from this device and sending the reply, the CMS might send it the following RQNT:

```
(1)

RQNT 1000 aaln/1@my.mta.com
MGCP 1.0

N: ca@the.callagent.
com:2427

X: 1234ab

R: hd(N)

S:
```

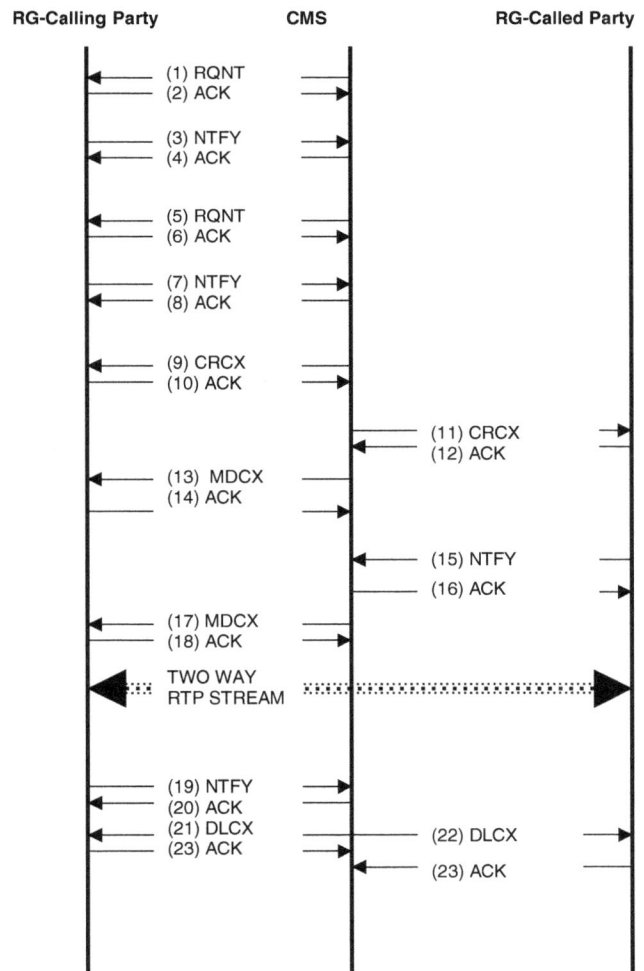

Figure 3.7: MGCP Signaling

With this message, the CMS is saying:

- *N*: When you see the events asked for in the R: line below, send them to the address *ca@the.callagent.com*, UDP port 2427.

- *X*: Use the following request ID in your response.

- *R*: Look for hook-detect event (i.e., look for when the user picks up the phone) and notify me right away when you see this.

- *S*: Don't play any signals.

MGCP signals and events are grouped into packages such as basic telephony, DTMF digits, analog line, etc. An MGCP package defines the text names for the events and signals, and their behavior. Signals are actions to be taken by the endpoint (such as play a tone), while events are state observations that the endpoint should report (such as on or off-hook). In some cases, events and signals can have associated parameters. For example, the caller id signal (ci) will be sent with the calling party name and number.

```
(2)
200 1000 OK
```

Continuing with the example, the calling party RG would respond with a positive acknowledgment (assuming that the endpoint was on hook).

```
(3)
NTFY 2000 aaln/1@my.mta.com MGCP 1.0
N: ca@the.callagent.com:2427
X: 1234ab
O: hd
```

Some time later, the user might pick up the phone to make a call. The calling party RG would detect this and report it to the CMS using an NTFY message: The O: line is used to report events that have been detected, in this case off hook.

```
(4)
200 2000 OK
```

The CMS would acknowledge that he received the message by sending:

```
(5)
RQNT 1001 aaln/1@my.mta.com MGCP 1.0
N: ca@the.callagent.com:2427
```

```
X: 1234ac
R: hu(N),[0-9](D)
S: dl
D: (XXXXXXXX|0T|00T)
```

The CMS would then send a new RQNT message to the calling party RG, telling it what to do next. A typical RQNT might look like the one above. With this message, the CMS is saying:

- *R*: Look now for a hang-up event (hu), and for dialled digits ([0–9]). In the case of hang-up, the CMS wants to be told right away (the meaning of the "*N*" parameter). In the case of digits, the CMS wants these to be accumulated according to the digit map below (*D*:) before being notified.

- *S*: Play dial tone (dl) to the user's handset.

- *D*: Match digits to this digit-map string. A digit map is a set of matching rules. In this case, the map is saying match any 9 digits, or the digit 0 with a brief timeout, or the digits 0 0, with a brief timeout.

```
(6)
200 1001 OK
```

The calling party RG would acknowledge the message with a 200 OK and would play dial tone to the user.

```
(7)
NTFY 2001 aaln/1@my.mta.com MGCP 1.0
N: ca@the.callagent.com:2427
X: 1234ac
O: 3,0,1,5,5,5,1,2,1,2
```

The user would dial his call and when a digit map match occurred, the calling party RG would notify the CMS with the dialed string. MGCP has conventions for managing signals. Timed signals, such as dial tone, will play until a predefined time out, or until an event occurs (such as a digit being detected). There are mechanisms in the protocol to keep signals active if required.

```
(8)
200 2001 OK
```

The CMS now has enough information to route the call to the destination. It would first acknowledge the NTFY.

```
(9)

CRCX 1002 aaln/1@my.mta.com MGCP 1.0

L: p:20, a:PCMU, s:off, e:on

C: abcde1234

M: recvonly

X: 1234ad

R: hu(N), hf(N)

S:
```

Next, a typical CMS would use the CRCX message to set up the calling party RG's side of the media connection. A CRCX creates an MGCP call leg. With this message, the CMS is saying that he would like a Call Leg created and:

- *L*: Would like to use the following for the media connection of this call:

    - a: use the PCMU (G711 mu-law) codec

    - p: use 20 ms packetization period (i.e., produce a voice packet of G711 samples, every 20 ms)

    - s: do not use silence detection

    - e: perform echo cancellation.

- *C*: Use this string as a call identifier.

- *M*: Place the connection in recvonly mode, meaning be prepared to receive media with the above characteristics, but do not send any media yet.

- *X,R,S*: Look for hang-up or hook flash events and report them to me. Do not play any signals.

The local connection options (L:) line is used to give the CMS preferences for how the media should be transported. A CMS could give a list of codecs and let the RG choose based on its current capabilities. The mode (M:) parameter is used to allow the CMS explicit control over the media direction. For example, to place a call on hold, the CMS can set the mode parameter to *inactive*.

```
(10)

200 1002 OK

I: abcd1234
```

```
v=0
o=- 25678 753849 IN IP4 192.168.3.11
s=-
c=IN IP4 192.168.3.11
t=0 0
m=audio 30000 RTP/AVP 0
```

The calling party RG, upon receipt of this message, will set up his side of the media connection. He will respond with an acknowledgment message that will also contain a description of his media connection, using the Session Description Protocol (SDP). With this message the RG is acknowledging that the connection has been set up. He returns a connection identifier for the CMS to use in future when managing this call leg (an endpoint may have more than one call leg). The portion of the response after the I: line is the description of the connection using SDP. We will postpone the details of SDP until the SIP section below. The most important information in the SDP response is the IP address given in the c= line, and the description of the media itself, given in the m= line. In particular, the number 30000 refers to the UDP port that should be used for media. The IP address and port returned in the SDP define where the remote, called party should send media. The CMS, assuming he has determined the destination for the call is an RG that he controls, now needs to alert that destination of the incoming call. Note that, if the destination is not directly controlled by this CMS, a CMS–CMS signaling protocol (such as SIP) would be used to communicate with the appropriate CMS. We will discuss SIP more below.

```
(11)
CRCX 1003 aaln/1@your.mta.com MGCP 1.0
L: p:20, a:PCMU, s:off, e:on
C: abcde1234
M: sendrecv
X: 1234ad
R: hd(N)
S:rg

v=0
o=- 25678 753849 IN IP4 192.168.3.11
s=-
c=IN IP4 192.168.3.11
t=0 0
m=audio 30000 RTP/AVP 0
```

MGCP allows the setup of the called party (aaln/1@your.mta.com) side of the connection and called-party alerting to be done with one message (although the use of multiple messages is also permitted). For example, the CMS (again assuming that he directly controls the destination) could issue the following CRCX message to the called-party RG. This message will tell the destination RG to:

- Set up his side of the media connection (i.e., create a call leg), using the PCMU codec, to start sending media and to be prepared to receive media (M:sendrecv).

- Look for an off-hook event, and report this immediately (R: hd(N)).

- Ring the user's phone (S:rg).

Note that the message contains the SDP returned by the calling party in its CRCX response. The destination RG will use this to determine where to send its media.

```
(12)
200 1003 OK
I: def1234

v=0
o=- 1111 23245 IN IP4 192.168.3.12
s=-
c=IN IP4 192.168.3.12
t=0 0
m=audio 40000 RTP/AVP 0
```

The destination RG will acknowledge the CRCX; the response will contain its media-session description:

```
(13)
MDCX 1004 aaln/1@my.mta.com MGCP 1.0
I: abcd1234
M: recvonly
X: 1234ae
R: hu(N), hf(N)
S: rt

v=0
```

```
o=- 1111 23245 IN IP4 192.168.3.12
s=-
c=IN IP4 192.168.3.12
t=0 0
m=audio 40000 RTP/AVP 0
```

The CMS now knows that the destination is ready for media and is alerting the called party. It can now issue an MDCX message to the calling party RG to get it to perform the following actions:

- To play ringback tone to its user (S:rt)

- To give the calling party RG the SDP describing the destination RG's media session

- To ask again for on hook and hook flash events to be reported when seen (R: hu(N), hf(N))

Note that the MDCX contains the call-leg identifier (I:) returned earlier.

```
(14)
200 1004 OK
```

The Source RG will begin playing ringback tone and the CMS will acknowledge the message. No SDP is returned in this case as the media session is unchanged. However, the source RG now knows where to send its media.

```
(15)
NTFY 5001 aaln/1@your.mta.com MGCP 1.0
N: ca@the.callagent.com:2427
X: 1234ad
O: hd
```

Assuming the called party eventually picks up, the destination RG will detect an off-hook event and report it to the CMS.

The CMS now knows that both parties are ready to be involved in the call. He must acknowledge the NTFY to the remote RGW:

```
(16)
200 5001 OK
```

and must tell the source RG to stop playing ringback tone (*s:*), and to enable bidirectional media (*M:sendrecv*). Again, this can be done in MGCP through one message:

```
(17)
MDCX 1005 aaln/1@my.mta.com MGCP 1.0
I: abcd1234
M: sendrecv
X: 1234ae
R: hu(N), hf(N)
S:
```

The source RG will perform the actions and acknowledge the message. Two-way communication between the calling and called parties will now be present.

```
(18)
200 1005 OK
```

At some point, the call will be ended. One of the RGs will detect the on-hook event and report this to the CMS (19-20). The CMS will then issue a DLCX message to both RGs (20-24), and can issue a RQNT(25-26) to the remaining party with a command to play an off-hook-warning tone if necessary.

We have discussed only a simple call. However the MGCP primitives involved in the simple call can be used by the CMS to perform advanced features. The call state and logic to make use of these primitives resides entirely in the CMS. An MGCP RG will most likely not be aware that it is part of a forwarded call or a multiparty conference, for example.

MGCP does not explicitly define security—i.e., how the MGCP signaling messages will be protected—but can run over various security layers such as IPsec and TLS.

3.4.2 Megaco/H248

H248, or Megaco, is a centralized protocol very similar to MGCP. Like MGCP it has the concept of a call manager that is a central point of control for distributed-media endpoints. Megaco is the result of a joint effort between the IETF and ITU-T Study Group 16. Thus, Megaco is defined jointly by RGC 3015 and ITU recommendation H248.

Like MGCP, Megaco is also a text-based protocol, but uses a different syntax. Megaco can run over UDP as well as TCP. Megaco uses SDP as well to describe its media sessions.

Megaco has two basic components: *terminations* and *contexts*.

Terminations are media sources or sinks such as a T1 timeslot, an analog port on a MTA, or an RTP port. This later type of termination is known as ephemeral, in that it can be dynamically created or deleted as needed. Terminations have properties and can have event packages associated with them. So, like MGCP, a Megaco CMS can use the protocol to ask for events to be reported and for signals to be played.

Contexts are defined as a mixing of terminations. A simple VoIP call is a context where an analog endpoint termination and a (newly created) RTP termination are brought together. Contexts are more powerful, however. Bringing several RTP terminations into one context, for example, is a way to create a multiparty conference. A context is created when the first termination is added to it. It is deleted when the last termination is removed.

The list of Megaco methods is given in Table 3.4.

Table 3.4: Megaco Signaling Methods

Method	Purpose
Add	Add a termination to a context
Modify	Modify the properties (or events/signals) of a termination
Subtract	Remove a termination from a context
Move	Moves a termination from one context to another
Audit	Used to query a termination for capabilities and status
Service Change	Used by RGs to report when a termination is brought in or out of service. This is used as a way to register terminations or move terminations to the control of another CMS
Notify	Used by RGs to report events on a termination

While Megaco was originally envisioned as a signaling protocol for all sorts of VoIP architectures, it is mostly used as a means for CMSs to manage physically separated, large media gateways (such as those providing VoIP-to-PSTN) internetworking. It is rarely used to control VoIP end user terminals such as IP or WLAN IP phones, so we won't discuss it much further.

3.4.3 H323

H323 is the "granddaddy" of VoIP signaling protocols (originally standardized in 1996). It is defined by ITU-T and was designed for voice and other media such as video. H323 is a distributed protocol with the following architectural elements:

- Terminal device (endpoint)

- Gateway

- Gatekeeper

- Multipoint control unit (MCU)

The terminal device in H323 is the H323 endpoint.

H323 gateways interface non-H323 networks to the H323 network. The gateway is responsible for establishing connections.

The gatekeeper is a service element for H323. It is possible (but not necessarily practical) to make H323 calls without a gatekeeper. The gatekeeper is responsible for translating telephone numbers to IP addresses. It also can manage bandwidth and performs a registration function. Supplementary services (such as call transfer or forwarding) are typically managed by the H323 gatekeeper.

Finally, the MCU is an H323 component responsible for managing multipoint conferences.

H323 is an umbrella standard. The basic call-signaling component of H323 is defined by H225. H225 is heavily influenced by the Q931 telephony-signaling standard. This standard is used in ISDN and other conventional telephony architectures. H225 has two components: call-control signaling and registration/admission/status (referred to as RAS).

The call-control signaling portion of H225 uses Q.931 signaling messages to set up, maintain and disconnect calls. These messages are sent over TCP connections. The main Q.931 signaling message types are given in Table 3.5.

Table 3.5: Q.931 Message Types

H225 (Q.931) Message Types	Direction	Purpose
Setup	Caller → Called Party	Sent by calling party to initiate a call
Call proceeding	Called Party → Caller	Used to indicate that the call is proceeding (note: this may come from an intermediate call-handling unit and not necessarily the final destination)
Alerting	Called Party → Caller	Used to indicate that the called party is being alerted
Connect	Called Party → Caller	Used to indicate that the called party has accepted the call
Connect ACK	Caller → Called Party	Indicates that the caller has seen the connect and that the call is now fully established

When a gatekeeper is present, the RAS component of H225 is used over a permanent signaling channel between the H323 gateway and the gatekeeper. The list of key H225-RAS message types is given in Table 3.6.

Table 3.6: H225 RAS Message Types

Message Type	Purpose
Registration Request	Used by terminal or gateway to register with the gatekeeper
Admission Request	Used by terminal to request access to network
Bandwidth Request	Used by terminal to request bandwidth for a call
Disengage Request	Used to cancel calls in progress (either from gateway to terminal or terminal to gateway)
Info Request/ Info Response	Request for/Respond with status information

Unlike MGCP, Megaco and SIP, H225 messages are not text based. Instead, H225 messages are encoded in the same fashion as in Q931, as lists of type/length/value records. Each record defines an information element (IE).

H.245 defines the control protocol for multimedia communication that is used with H323. It is used for the exchange of messages between the two endpoints involved in the call. Once the call has been set up via the H225 signaling, H245 signaling takes over to negotiate the parameters to be used for the media. Some of the additional functions of H245 include:

- Master–slave assignment. With H323 one side of the connection is designated the master to avoid conflicts that might arise.

- Capability exchange, including what audio codecs are available.

- Establishing and managing logical channels to be used for the media transfer.

- Media changes during the call (for example, if one side wishes to switch codecs).

Note that H245 signaling does not use the Session Description Protocol.

H245 messages are encoded in ASN.1 syntax. The key H245 message types are given in Table 3.7.

Table 3.7: H245 Message Types

H245 Message Type	Purpose
Master/Slave Negotiation	Defines which terminal is the master and which is the slave for the duration of the call.
Terminal Capability Set	Set by a terminal to publish its media capabilities.
Send Terminal Capability Set	Used by a terminal to request that the remote terminal publish its capabilities.
Open Logical Channel	Opens a logical channel that is used to transport a media stream.
Close Logical Channel	Closes a logical channel.
Request Mode	A receive terminal uses this message to ask for a specific type of transmission such as audio or video.
End Session	Ends the H245 session.

A suite of H.450 series standards defines H323 supplementary services, including:

- call transfer
- call diversion
- call hold
- call waiting
- message-waiting indicator

Standard H.235 defines security and encryption protocols for H323.

3.4.4 Session Initiation Protocol (SIP)

The Session Initiation Protocol is the current leader in standards-based VoIP call-signaling protocols. Originally defined to invite users to multimedia conferences, SIP is a powerful protocol for managing VoIP calls. SIP is defined as a base protocol in RFC 3261. There are, however, numerous extensions to it, defined in subsequent RFCs. Like MGCP and Megaco, SIP is text-based. It can run over UDP and TCP, as well as transport layer security (TLS). Like H232, SIP can be considered a distributed signaling protocol.

SIP is based on HTTP and the Simple Mail Transport Protocol (SMTP). Like MGCP and Megaco, it uses SDP to actually describe the media sessions it is controlling.

SIP can be a direct, peer-to-peer protocol—e.g., between a calling and called party. The more common scenario is when intermediate SIP devices are present, known as SIP proxies. SIP proxies perform a variety of functions (in some way similar to what the H323 gatekeeper does). An SIP endpoint is referred to as an SIP user agent. The main signaling messages defined for SIP are listed in Table 3.8.

Table 3.8: SIP Signaling Methods

SIP Method	Direction	Purpose
Invite	Caller → Called Party	Used to initiate a call.
Update	Either	Used to refresh the media session description to be used during the call.
Message	Either	Used as an ad-hoc mechanism to exchange information such as the number of messages waiting in a user's voice mail queue. Also used to convey instant messaging.
Subscribe	Subscribing Entity → Notifying Entity	Used to allow a SIP entity to register for events.
Notify	Notifying Entity → Subscriber	Used to report events.
Publish	Any	Similar to *Notify*.
Cancel	Any	Used to cancel a pending request.
ACK	Caller → Called Party	Used to acknowledge that the *200 OK* response to the Invite has been received (and that the call is fully established).
Info	Any	Used to carry out-of band signaling between SIP entities. For example, info can be used to send DTMF digits that have been collected during the call (e.g., to communicate with an IVR system).
Register	SIP user agent to proxy	Used to register a SIP user agent to a proxy.
Refer	Any	Used to request that the recipient SIP entity issue a new SIP request to another SIP entity (e.g., for call transfer).
PRACK	Caller → Called Party	Used by caller to acknowledge intermediate responses to the Invite.
Bye	Either	Used to terminate a call.
Transaction Responses	Any	Used to send acknowledgment codes for received messages, e.g., *100 Trying* or *200 OK* can be sent in response to an *Invite*.

Let's look at a simple user-agent-to-user-agent SIP call (with a single proxy). The call flow is shown in Figure 3.8.

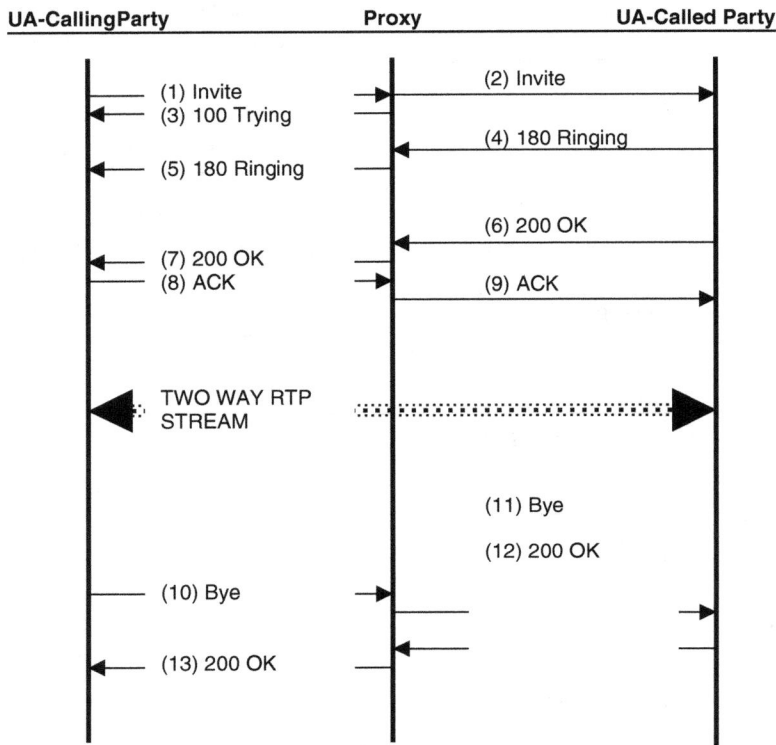

Figure 3.8: SIP Call Establishment

Assume the UA-Calling Party is named *user1@the.sipnet.com* (in SIP, identifiers are known as URI or universal resource identifiers) and the UA-Called Party is named *user2@the.sipnet. com,* and that the proxy is named *proxy@the.sipnet.com.* We start with the calling party issuing an Invite message to the proxy.

(1)

```
INVITE sip:user2@the.sipnet.com SIP/2.0

Via: SIP/2.0/TCP user1.the.sipnet.com:5060;branch=abcd12345Dc

Max-Forwards: 70

Route: <sip:proxy.the.sipnet.com;lr>

From: user1 <sip:user1@the.sipnet.com>;tag=845rfgt35

To:   user2 <sip:user2@the.sipnet.com>
```

```
Call-ID: 1256873425256

CSeq: 1 INVITE

Contact: <sip:user1@the.sipnet.com ;transport=tcp>

Content-Type: application/sdp

Content-Length: 151


v=0

o=user1 123456 123456 IN IP4 user1.the.sipnet.com

s=-

c=IN IP4 158.218.1.150

t=0 0

m=audio 30000 RTP/AVP 0

a=rtpmap:0 PCMU/8000
```

The Invite message contains two sections: the message headers and the session description, encoded using the Session Description Protocol (SDP). SIP utilizes many types of message headers. A partial list is given in Table 3.9. Notice that the first line of the message identifies the called-party destination. The SIP protocol is unconcerned with the call control steps necessary to obtain the destination (i.e., playing dial tone, collecting digits, entering the destination user name in a call menu, or selecting from an address box). Contrast this to the MGCP protocol, where a portion of the protocol message exchange is involved in controlling these steps.

Table 3.9: SIP Header Description

SIP Header	Purpose	Comment
Via	List the SIP Identifier of all SIP devices that touch the message	Proxies will use this header to identify themselves
To	Gives the URI of the destination for the message	
From	Gives the URI of the source of this message	
Contact	Gives the URI of the ultimate source of this message	Typically the same as the From Header
Call-ID	Opaque call identifier, set by calling party	
CSeq	Transaction sequence number	
Content-Type	Defines the type of information contained in the message body (e.g., SDP)	e.g.: application/sdp if remainder of message is SDP information
Content-Length	Gives the length of the information in the message body	Length of the rest of the message
Max-Forwards	Maximum number of SIP routing hops that this messages should visit	Decremented by one during each hop

SDP (RFC 2327) is a text-based encoding method for defining multimedia sessions. For SIP, and many VOIP signaling protocols, SDP is used to define the types of audio media stream(s) that the calling party wishes to use and how/where they should be sent. In this example, the Invite is saying that user1 is prepared to receive RTP encapsulated voice samples encoded with the PCMU (mu-law G.711) codec, at IP address 158.218.1.150, UDP port 30000. A summary of the important SDP parameters is given in Table 3.10.

Table 3.10: Session Description Protocol Parameters

SDP Parameter	Purpose	Comment
v=	Version (0,1,2,..)	Set to 0 initially, but incremented on subsequent SDP exchanges if SDP changes during a call.
o=<username> <session id> <version> <network type> <address type> <address>	Session origin information.	
t= <start> <end>	Session start and end time	Typically 0 0 for VoIP.
s=	Session name	Not used in VoIP sessions typically.
c=	Connection information	For VoIP, typically: **IN IP4 *ipaddr***
m= <media> <port>/<number of ports> <transport> <fmt list>	Defines one media session. <fmt list> is a list of RTP payload types that the receiver is prepared to use for the call. Dynamic payload types must be mapped to specific media codecs using the **rtpmap** attribute.	• <media> is typically set to ***audio*** for VOIP calls. • <transport> is typically ***RTP/AVP*** for VOIP calls. • <port> is the UDP port where media should be sent to. More than one media type can be used for a call. In this case, the <fmt list> will contain more than one value.
a= <attribute> <value>	Defines an attribute.	
a= rtpmap: <pl> <media>	Maps a payload type to a specific codec. This is required when a dynamic payload type is used instead of an industry-assigned one.	
a= fmtp: <string>	Used to provide codec-specific information.	
a= recvonly,inactive, sendonly, sendrecv	Used to specify media direction control.	e.g., **a=inactive** is used to place a call on hold.
a= ptime:<value>	Used to define the desired voice packetization period to use for the media stream.	

The proxy then forwards the received Invite to the called party. In a real scenario, the Invite might be forwarded to another proxy instead. The Invite is changed slightly. The Proxy will add a new VIA header to indicate that it is involved in the call, and can issue a Record-Route header to make sure that it see replies. It may add a "received=" to indicate the IP address from which it received the Invite; this is useful in NAT transversal, for example. The Max-Forwards value is also decremented.

(2)

```
INVITE sip:user2@the.sipnet.com SIP/2.0

Via: SIP/2.0/TCP proxy.the.sipnet.com:5060;branch=abcd1236721

Via: SIP/2.0/TCP user1.the.sipnet.com:5060;branch=abcd12345Dc
    ;received=158.218.1.150

Max-Forwards: 69

Record-Route: <sip:proxy.the.sipnet.com;lr>

Route: <sip:proxy.the.sipnet.com;lr>

From: user1 <sip:user1@the.sipnet.com>;tag=845rfgt35

To:  user2 <sip:user2@the.sipnet.com>

Call-ID: 1256873425256

CSeq: 1 INVITE

Contact: <sip:user1@the.sipnet.com ;transport=tcp>

Content-Type: application/sdp

Content-Length: 151


v=0

o=user1 123456 123456 IN IP4 user1.the.sipnet.com

s=-

c=IN IP4 158.218.1.150

t=0 0

m=audio 30000 RTP/AVP 0

a=rtpmap:0 PCMU/8000
```

The proxy will also send a provisional acknowledgment to the calling party to indicate that it has received the initial Invite and is proceeding with the call:

(3)

```
SIP/2.0 100 Trying
Via: SIP/2.0/TCP user1.the.sipnet.com:5060;branch=abcd12345Dc
    ;received=158.218.1.150
From: user1 <sip:user1@the.sipnet.com>;tag=845rfgt35
To:  user2 <sip:user2@the.sipnet.com>
Call-ID: 1256873425256
CSeq: 1 INVITE
Contact-Length: 0
```

The called party, upon receiving the Invite, will take some action to alert its user of the incoming call. Again, unlike MGCP, SIP does not concern itself with this aspect of call control. The called-party user agent must manage the alerting mechanism (i.e., ringing the phone, playing a musical ring tone, etc.) on its own. The called party, however, does return a provisional acknowledgment to the proxy indicate that this alerting is taking place. The proxy will then forward this to the calling party:

(4), (5)

```
SIP/2.0 180 Ringing
Via: SIP/2.0/TCP proxy.the.sipnet.com:5060;branch=abcd1236721
Via: SIP/2.0/TCP user1.the.sipnet.com:5060;branch=abcd12345Dc
    ;received=158.218.1.150
Record-Route: <sip:proxy.the.sipnet.com;lr>
From: user1 <sip:user1@the.sipnet.com>;tag=845rfgt35
To:  user2 <sip:user2@the.sipnet.com>;tag=675432
Call-ID: 1256873425256
Contact: <sip:user2@the.sipnet.com ;transport=tcp>
CSeq: 1 INVITE
Contact-Length: 0
```

When the called party picks up the call, the called-party user agent will send a final response indicating that the call has been established. In this response, he will return an SDP description of his side of the call. This will be sent to the proxy, who will forward it to the calling party:

(6), (7)

```
SIP/2.0 200 OK
Via: SIP/2.0/TCP proxy.the.sipnet.com:5060;branch=abcd1236721
Via: SIP/2.0/TCP user1.the.sipnet.com:5060;branch=abcd12345Dc
    ;received=158.218.1.150
Record-Route: <sip:proxy.the.sipnet.com;lr>
From: user1 <sip:user1@the.sipnet.com>;tag=845rfgt35
To:  user2 <sip:user2@the.sipnet.com>;tag=675432
Call-ID: 1256873425256
Contact: <sip:user2@the.sipnet.com ;transport=tcp>
CSeq: 1 INVITE
Content-Type: application/sdp
Content-Length: 151

v=0
o=user2 654321 654321 IN IP4 user2.the.sipnet.com
s=-
c=IN IP4 158.218.1.151
t=0 0
m=audio 40000 RTP/AVP 0
a=rtpmap:0 PCMU/8000
```

The calling party, upon receipt of the 200 OK, now can confirm the media format(s) to be used for the call, and where exactly to send his RTP packets. Two-way conversation can now take place. It remains for him to acknowledge that he has seen the 200 OK by sending a final SIP message, ACK. Again, this message is sent first to the proxy, who then forwards it to the called party:

(8), (9)

```
ACK sip:user2@the.sipnet.com SIP/2.0
Via: SIP/2.0/TCP proxy.the.sipnet.com:5060;branch=abcd1236721
Route: <sip:proxy.the.sipnet.com;lr>
```

```
From: user1 <sip:user1@the.sipnet.com>;tag=845rfgt35

To:  user2 <sip:user2@the.sipnet.com>;tag=675432

Call-ID: 1256873425256

Contact: <sip:user2@the.sipnet.com ;transport=tcp>

CSeq: 1 ACK

Contact-Length: 0
```

When one side wishes to terminate the call, it will send a BYE message to the other side (steps 10, 11). The other side will respond with a 200 OK (steps 12, 13), and the call will be torn down.

3.5 Voice-over-IP Media

We have discussed the mechanisms used in VoIP signaling. Now let's dive into the media path.

As mentioned above, the majority of open-standard-based VoIP implementations use the Real-Time Transport Protocol (RTP) for the media path. RTP is defined in RFC 3550. A media stream that is RTP encapsulated will feature the following:

- An RTP header prepended to each media packet.

- A separate stream of control packets, called the Real-Time Control Protocol (RTCP).

The RTP header is illustrated in Figure 3.9.

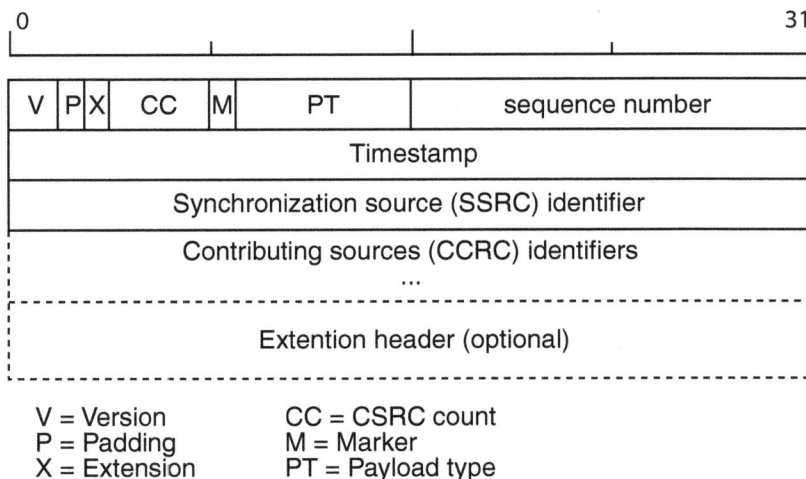

```
 0                                                          31

 V P X  CC  M   PT           sequence number

               Timestamp

        Synchronization source (SSRC) identifier

        Contributing sources (CCRC) identifiers
                          ...

              Extention header (optional)


 V = Version      CC = CSRC count
 P = Padding      M = Marker
 X = Extension    PT = Payload type
```

Figure 3.9: RTP Header

The RTP header is 12 bytes in length plus zero or more optional 4-byte contributing source-identifier blocks. The first two bits of the header define the RTP version (V), which is currently 2.

The *P* bit indicates if the RTP packet contains padding (to get the packet aligned on 32-bit boundaries). If the *P* bit is set, then one or more padding bytes are present at the end of the packet and the very last byte of the packet will be a count of their number. Padding is sometimes required for some encryption algorithms, for example.

The *X* bit, when set, indicates that an RTP extension header follows the main header. RTP extension headers can be used for a variety of purposes, but are normally not present in RTP encapsulated voice media streams.

The 4-bit *CC* field gives a count of the number of optional contributing source identifiers. For a normal point-to-point call, *CC* is always 0, and there are no contributing sources. Contributing sources are used mostly with large, multiparticipant conferences.

The usage of the *M* bit is determined by the RTP payload. It is typically used to indicate some event in the real-time flow. For voice, the *M* bit is usually set on the first voice packet of a talk spurt.

The 7-bit payload field is used to identify the type of media being carried in the payload. As we discussed above with VoIP call signaling, one or more voice codecs can be selected for use in a call. Each will have an associated payload type that is carried in this field. For calls described through the SDP, the payload types associated with a call are defined in the *m=* and *a=rtpmap* lines. This way, a receiver can figure out how to process each payload correctly.

The 16-bit sequence number field is incremented by one for each RTP packet sent. The RTP receiver can use this field to detect gaps in the media stream.

The 32 bit timestamp gives the sampling time associated with the first sample in the payload. Its units are the sampling rate used for the media. For narrowband voice, the sampling rate is 8 kHz, so that an increment of one RTP timestamp corresponds to 125 μs. If voice packets contain 20 ms of samples, for example, we would expect to see the timestamp field increment by 20 ms/ 125 μs or 160 in each subsequent RTP packet.

The 32-bit synchronizing source (SSRC) field is a random identifier selected by the transmitter and is used to identify the media session. An RTP receiver could use this field, for example, to route payloads to the correct destination if he uses one UDP port for all media streams.

RTCP is a separate control protocol defined for RTP. While designed to facilitate large multiuser, multimedia conferences, components of this protocol are applicable to a typical point-to-point VoIP call. RTCP packets are exchanged periodically (over UDP) between the two endpoints of a call. By convention, these packets are sent to the UDP port number that is one above the UDP port signaled for the media stream (RTP ports are constrained, therefore, by convention to be even).

The major types of RTCP messages (blocks) are given in Table 3.11.

Table 3.11: RTCP Block Types

RTCP Block Type	Purpose
Sender Reports	Contains sending-side statistics such as the number of RTP packets and octets that have been sent.
Receiver Reports	Contains receiving-side statistics such as the number of RTP packets that have been received, the number of packets that have been detected as lost by the network (e.g., sequence number gaps), and an average of the receive jitter that is seen.
Bye	Used to signal the end of media stream.
Session Description (SDES)	Contains session-level description information such as the name of the session. Typically set to random values in VoIP calls to protect caller privacy.
RTCP-XR • VOIP report	These are extension blocks that carry more detailed session quality information. Of interest to VOIP is the VoIP report subblock that contains a voice quality score. Other XR subblocks include run length encoded logs of packet losses (gaps), packet duplicates and arrival times.

3.6 The Overall Picture

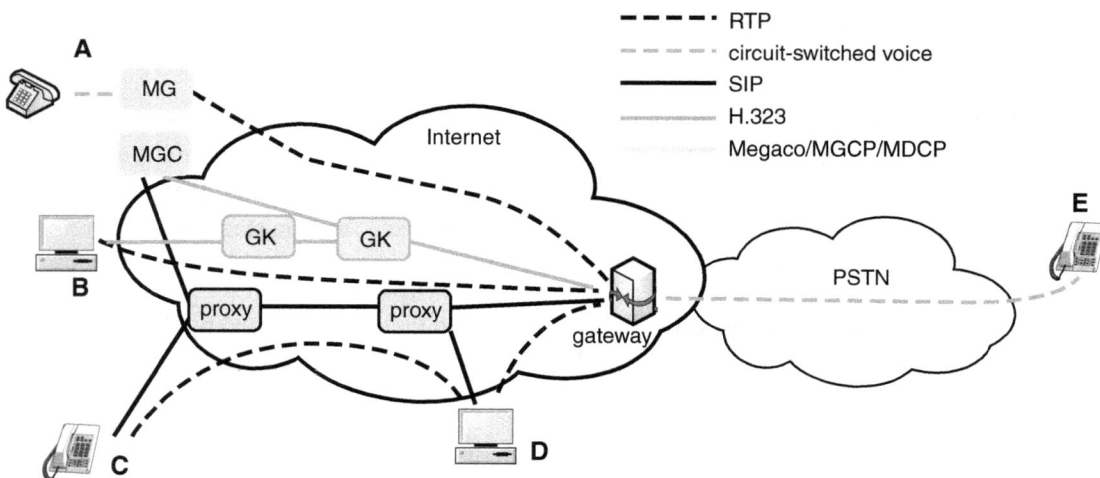

Figure 3.10: VoIP in the Real World

Now that we have the details of how VoIP is implemented, it is instructive to step back and take a bird's eye view of the whole picture. Figures 3.10 and Figures 3.11 are detailed representations of Figure 3.5. The reader is encouraged to ponder over these figures for a few minutes before reading ahead. The book so far should have helped in understanding these figures completely.

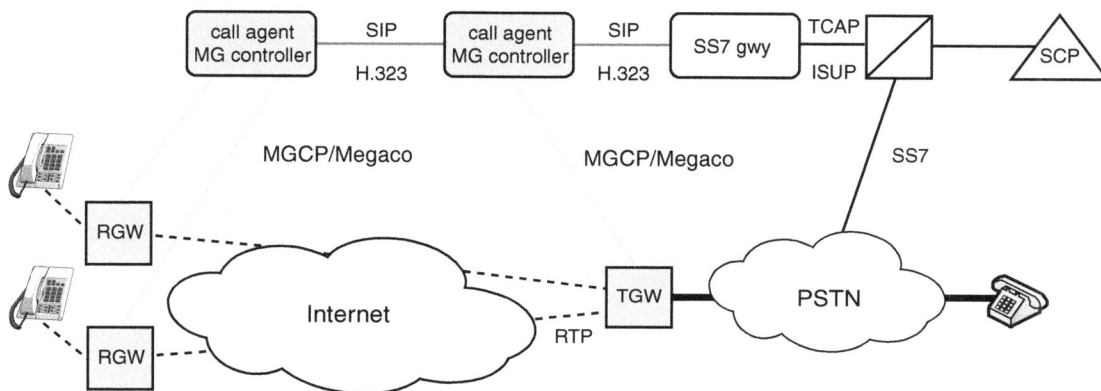

Figure 3.11: VoIP Protocols and Network Elements

References

[1] RFC 3435 Media Gateway Control Protocol (MGCP) Version 1.0, F. Andreasen, et al., January 2003.

[2] Packet Cable™ 1.5 Specifications: Network-Based Call Signalling Protocol, PKT-SP-NCS1.5-IO2-050812, Cable Television Laboratories, Inc. 2004–2005.

[3] RFC 2327 – SDP: Session Description Protocol, Handley, M. and V. Jacobson, April 1998.

[4] RFC 3015—Megaco Protocol Version 1.0, F. Cuervo et al., November 2000.

[5] RFC 3261—SIP: Session Initiation Protocol, J Rosenberg et al., June 2002.

[6] RFC 2833—RTP Payload for DTMF Digits, Telephony Tones and Telephony Signals, H. Schulzrinne et al., May 2000.

[7] RFC 2198—RTP Payload for Redundant Audio Data, C. Perkins et al., September 1997.

[8] RFC 3550—RTP, a Transport Protocol for Real-Time Applications, H. Schulzrinne et al., July 2003.

Wireless Local Area Networks

4.1 Introduction

With the phenomenal growth in data traffic (think Internet), there has been a demand for wireless networks capable of transferring data traffic along with voice traffic. Just as in the wired world, the field of wireless is seeing the integration of voice and data networks. Second-generation (2G) wireless networks, currently the most widely deployed and used, have been enhanced to support data. Such networks are sometimes referred to as 2.5G in order to distinguish them from voice-only 2G wireless networks. Moreover, 3G networks, the next-generation wireless networks, have been designed with inherent support to carry both voice and data.

Given the capabilities of 2.5G and 3G to carry data, it may not be apparent at first why there was a need to design another wireless standard. As we will see, 802.11 and 3G are more different than they are similar. Yes, both of them are wireless network standards and, yes, both of them support both voice and data, but 802.11 is a LAN standard meant to connect wireless clients in a small geographical area whereas 3G aims to provide wide-area (universal) wireless connectivity.

The first widely deployed wireless data network standard has been IEEE's 802.11 standard. The 802.11 standard is a suite of protocols defining an Ethernet-like communication channel using radios instead of wires. Such networks are referred to as *WLANs (wireless local area networks)* and the technology is more popularly referred to as *Wi-Fi*. WLANs allow users to connect to a network (and by extension to the Internet) without the wires. Put simply, 802.11 is Ethernet (802.3) without the wires. Just as we use 802.3 to form wired local area networks (LANs), we can use 802.11 to create WLANs. On the positive side, since there are no wires to lay down to create the network, setting up WLANs is much easier than setting up LANs. On the other hand, due to the nature of the wireless medium, the packet loss experienced in WLANs is much more than that in wired LANs.

Another distinguishing feature of the 802.11 standard is that it operates in the unlicensed frequency spectrum. This means that 802.11 service providers (popularly referred to as WLAN service providers) do not have to pay a "spectrum-usage" fee to their governments. Contrast this with 3G where service providers have spent billions of dollars in purchasing the 3G spectrum, which was auctioned by governments worldwide just a few years ago. Operating in the unlicensed frequency spectrum has the advantage of keeping operating costs low but also means no protection from interference caused by other users. This makes the wireless operating environment even more difficult to operate in.

4.2 The Alphabet Soup

The 802.11 standard specifies protocols for the PHY and the MAC layers of the OSI stack. However, the sheer number of amendments/enhancements to the base 802.11 standard is often a source of confusion. In this section we clarify the alphabet soup of 802.11 a, b, e, g, h, i, j, k and the list keeps growing ...

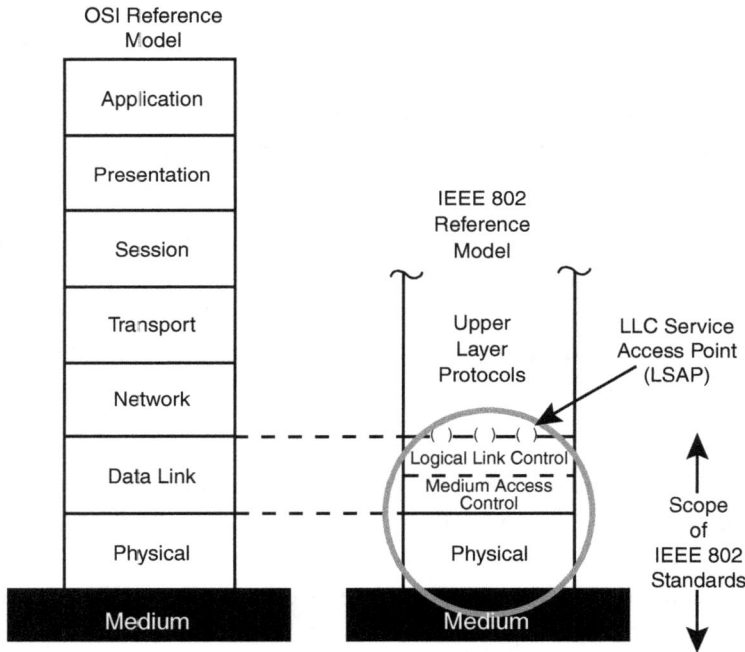

Figure 4.1: 802.x Mapping to the Seven-Layer OSI Model

The base 802.11 standard was ratified in 1997. It specified a unified MAC layer and three separate PHY layers (DSSS, FHSS and Infrared) that provided for data rates of 1 to 2 Mbps. The IEEE formed various Task Groups to enhance this base standard. These enhancements concentrated on various facets of the standard.

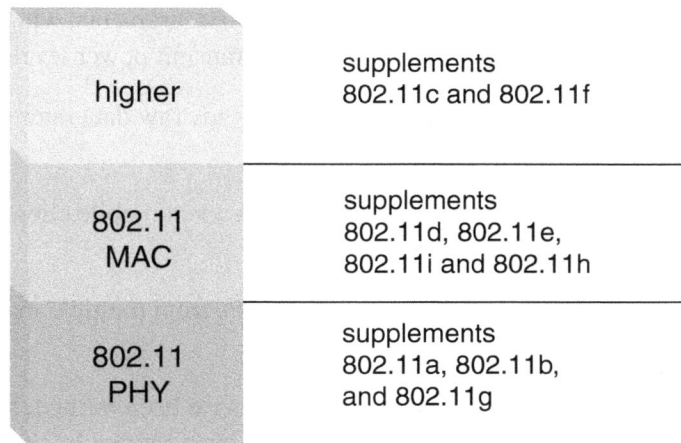

Figure 4.2: 802.11 Layers and Specifications

802.11a: PHY enhancement operating in the 5-GHz band uses OFDM modulation to provide raw data rates up to 54 Mbps.

802.11b: PHY enhancement operating in the 2.4-GHz band uses CCK modulation to provide raw data rates up to 11 Mbps.

802.11c: Provides required information to ensure proper bridge operations.

802.11d: MAC and PHY enhancements to allow 802.11 to operate in regulatory domain of various countries.

802.11e: MAC enhancements to provide Quality of Service for real-time applications like voice and video.

802.11f: Recommended practices for Inter-Access Point communication.

802.11g: PHY enhancement operating in the 2.4-GHz band uses CCK or OFDM modulation to provide raw data rates up to 54 Mbps.

802.11h: MAC and PHY enhancements to 802.11a to help with regulatory compliance in various countries.

802.11i: MAC enhancement to improve security mechanisms by exploiting existing standards like AES, 802.1X, RADIUS.

802.11j: Regulatory extensions to allow for operation in the 4.9-GHz and 5-GHz bands in Japan.

802.11k: MAC layer enhancements to standardize radio measurements, which are useful for making better decisions regarding frequency reuse, transmit power levels, etc.

802.11n: PHY and (possibly) MAC enhancements to provide raw data rates greater than 150 Mbps. Standard still under discussion.

802.11r: MAC enhancements to reduce handoff latencies when mobile clients transition between access points or cells in an ESS.

802.11s: Infrastructure mesh standard to allow 802.11 APs from multiple manufacturers to self-configure into multihop wireless topologies.

It is important to note that not all of the above standards have been ratified. Many of them (like k, n, r and s) are still under discussion and have not been ratified by the IEEE.

4.3 Network Architecture

The 802.11 standard specifies two network architectures: Infrastructure BSS and Independent BSS. In an Independent BSS (that is, IBSS), 802.11 stations communicate directly with each other. This architecture was targeted towards ad hoc networking where a small number of stations can randomly come together (within each other's radio range) and form a network. For the purpose of this book, we discuss Infrastructure BSS only and VoWLAN (or VoWi-Fi) is intended to mean voice-over-Infrastructure BSS.

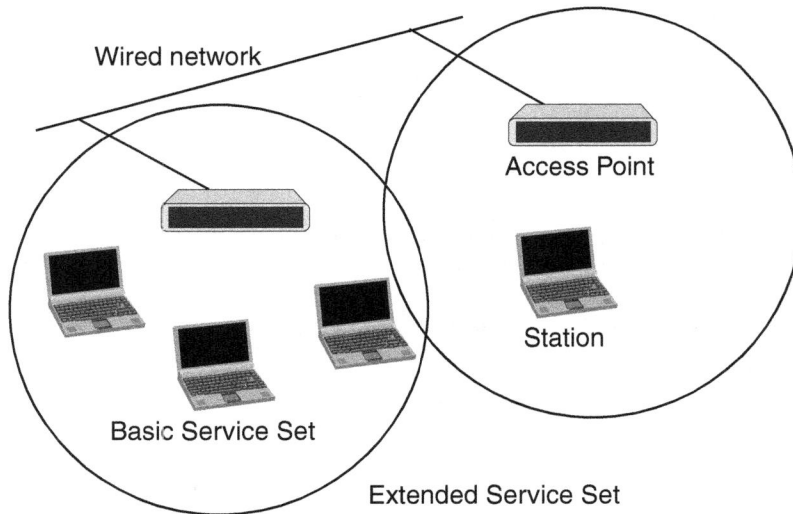

Figure 4.3: 802.11 Service Sets

A typical 802.11 network consists of four major physical components. First, we have the *station* (or STA). A station is an endpoint of connection with a wireless interface used to access the 802.11 network. Typical examples of stations are laptops, palmtops and other hand-held computers. Figure 4.3 shows laptops as stations.

Second, we have the *access point* (AP). An access point is basically a layer-2 bridge that has one wireless interface and one wired interface. It is therefore the AP that connects the wireless LAN (or rather the stations in the WLAN) to the wired LAN. It is a law of physics that radio propagation effects limit the range of wireless transmissions. In effect, this means that the geographical range served by the base station[1] is limited. This range can be increased by increasing the transmission power level at the base station. However, the cellular concept requires (and exploits) that the transmission range of a base station be limited. 802.11 has an additional constraint: it operates in an unlicensed band in the spectrum. By law, the transmission power level in the unlicensed band is restricted. This restricts the range of an access point in 802.11 networks. This area is called the *Basic Service Area* (BSA). Figure 4.3 shows the BSA of an AP as encircled. While a station is within the range of an AP, it has access to the wired network and other stations in this BSA.[2] The set of stations within a BSA that can communicate with each other is called the *Basic Service Set* (BSS).

Third, we have the wireless medium that actually carries the data between the stations and the AP. The use of radio waves to carry data significantly complicates the design of the physical layer since the wireless medium presents a much bigger set of challenges than any other medium. To deal with multiple physical layers, all of which use the single 802.11 MAC layer, the 802.11 standard splits the physical layer into two components: the PMD and the PLCP. The *Physical Medium Dependent* (PMD) is responsible for actually transmitting the frames onto the wireless medium. As is obvious from the name, this layer is different for each physical layer (DSSS, OFDM, FHSS, etc.). The *Physical Layer Convergence Protocol* (PLCP) is responsible for providing a uniform interface of the various physical layers (and the PMDs) to the 802.11 MAC layer. The position of the PLCP in the OSI model is hazy. The PLCP sits between layers 1 and 2 and abstracts the variations of the physical layers so that the 802.11 MAC can function independently of the physical layer in use.

Finally, we have the *Distribution System* (DS). The DS refers to the wired network that the AP connects to on its wired interface. When a packet/frame arrives over the wireless interface at the AP, the AP forwards it on its wired interface to the DS. The DS is responsible for delivering it to the right node, which may be a station on the wired network, another AP or a router. Also, if the station is mobile and if it moves out of the range of the AP and enters into

---

[1] BTS in GSM; Node-B in 3G, etc.

[2] 802.11 infrastructure networks require even the interstation communication in a BSA to go through the AP. 802.11e does allow stations to bypass the AP for communicating with each other.

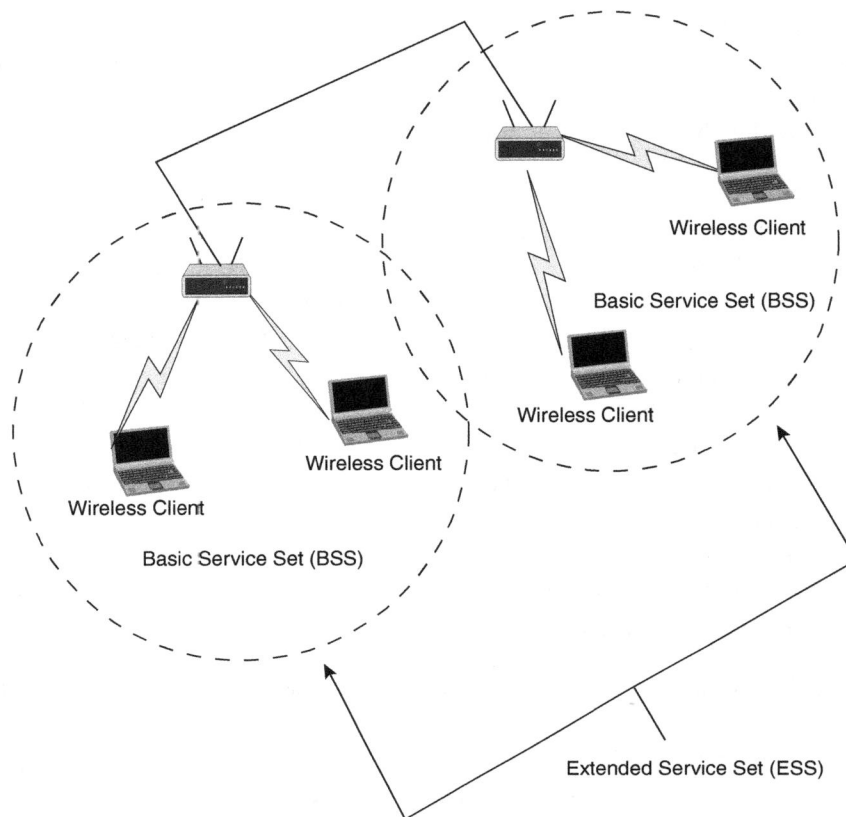

Figure 4.4: 802.11 Distribution System

the range of another AP, this station expects its session to be uninterrupted. Obviously, this requires that the two APs be able to communicate with each other. The APs communicate with each other using the DS. In other words, the DS connects various APs to form an *Extended Service Set* (ESS). The existence of a DS (and hence the existence of an ESS) allows for the possibility of transparent hand-off when a station is mobile. The 802.11 standard does not specify any particular technology for the distribution systems. However, most commercial implementations of 802.11 use Ethernet as the distribution system.

4.3.1 Connection Setup

In 802.11, connections/associations work at the link layer and an 802.11 station that is associated or connected with an access point may or may not be actively receiving or transmitting data. All this "association" means is that the station can transmit/receive data using this AP. Figure 4.5 shows the basic 802.11 connection-setup process. The following sections describe the process in some detail.

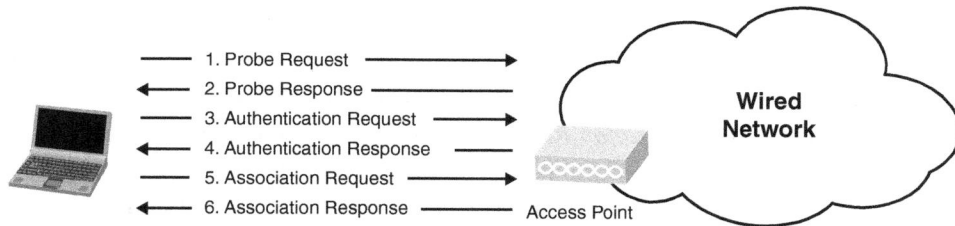

Figure 4.5: 802.11 Connection Setup

4.3.1.1 Scanning

When a station wishes to connect to an 802.11 network, the first thing it needs to do is to find a network. This is accomplished by the scanning process. Scanning comes in two basic flavors: passive and active. In passive scanning, a station listens (waits passively) for *Beacons*. In active scanning, the station broadcasts *Probe Request* messages on all (or a subset of all) channels to elicit responses from available APs. The *Probe Request* message contains the SSID of the network (ESS) that this station wishes to connect to.[3] An AP that receives the *Probe Request* message may reply back with a *Probe Response* message if it wants to allow this station to connect to its network. At the end of the scanning process, the station has a list of all wireless networks it has access to. This information is usually maintained in the form of a Site Table by the station and the station can use any algorithm to choose the network it wishes to associate with. We will discuss this in more detail in Chapter 8. Once this choice is made, the next step is for the station to authenticate itself to the AP.

4.3.1.2 Authentication

In effect, the access point serves as a gatekeeper to the network, deciding which stations should be allowed to access the network. For this purpose, the access point may use several rules. For example, it may allow only those stations to connect that explicitly specified the correct SSID in the *Probe Request* message or it may allow stations with only certain MAC addresses or it may use the 802.11 authentication process for this purpose.

802.11 specifies two authentication processes: OSA (Open System Authentication) and SKA (Shared Key Authentication). At a very high level, the authentication process consists of *Authentication Request*(s) and *Authentication Response*(s) exchanged between the station and the AP to carry out authentication. We look at these authentication processes in much more detail in Chapter 7.

---

[3] The SSID field may be left blank in order to indicate that the station wishes to connect to any network that it finds.

4.3.1.3 Association

The final step in the process is association. The aim of the association process is to establish a logical connection between the station and the access point. The association process starts with the station sending an *Association Request* to the access point. This request contains parameters such as the capability information of the station and the rates that the station can support. The access point responds with an *Association Response* message that may accept or reject the association depending on the parameters provided in the *Association Request* message. In case of a successful *Association Response* message, the AP also includes an *Association ID* (AID) to uniquely identify this association. Once a station is associated with the access point (and therefore the BSS), the network now knows "the location" of this station. This allows the network to deliver data sent to the station. This is so because only after registering with the AP can the DS (distribution system) know that a particular mobile node is being served by one of the APs in an ESS. Since traffic can now flow bidirectionally between the station and the network, the link-layer connection is now established at the completion of the association process.

4.4 802.11 Framing

Now that we have an overall idea of the 802.11 network architecture, it is useful to see what an 802.11 frame looks like. Figure 4.6 shows the generic 802.11 frame without the Physical Layer headers (which are discussed in section 4.6). We discuss each field in the following sections.

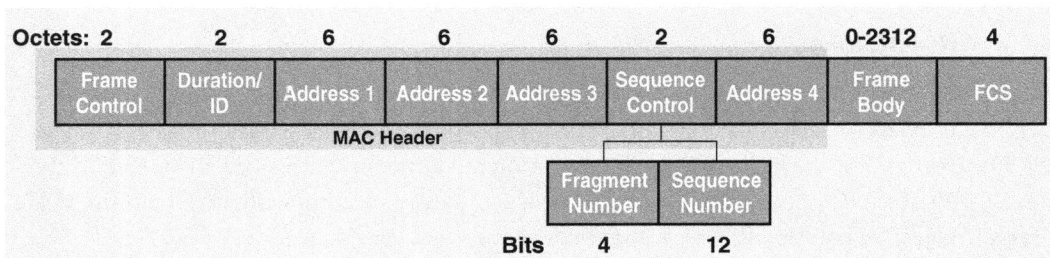

Figure 4.6: 802.11 Frame Format

4.4.1 Frame Control

The most interesting field is the Frame Control field and Figure 4.7 shows the detailed break-up of this field. We discuss the purpose of each of these fields below.

bits: 2	2	4	1	1	1	1	1	1	1	1
protocol	type	subtype	to DS	from DS	more frag	re-try	pwr mng	more data	WEP	order

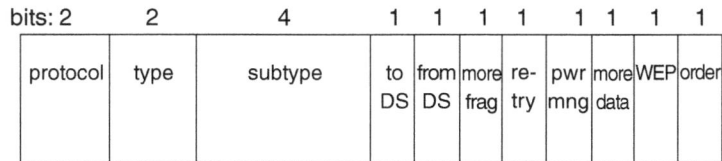

Figure 4.7: 802.11 Frame Control Field

Protocol: Version of the 802.11 MAC in the rest of the frame. 0 for base 802.11 standard.

Type and Subtype: The type and subtype fields together uniquely identify the purpose of each frame. There are currently three types of frames defined: Management Frames (Type 0), Control Frames (Type 1) and Data Frames (Type 2). Type 3 is reserved. Within Each type of frame, there are multiple subtypes defined. Tables 4.1 through 4.4 give a brief explanation. All values in the tables are in binary format.

Table 4.1: Type 00: Management

Subtype	Name	Explanation
0000	Association Request	See section 4.3.1.3
0001	Association Response	See section 4.3.1.3
0010	Reassociation Request	Similar to Association Request but used when station was previously connected to another AP in the ESS (Facilitate IAPP)
0011	Reassociation Response	Similar to Association Response; sent in response to a Reassociation Request
0100	Probe Request	See section 4.3.1.1
0101	Probe Response	See section 4.3.1.1
1000	Beacon	See section 4.3.1.1
1001	Announcement Traffic Indication Map	Used for IBSS
1010	Disassociation	Used by AP / station to end an association
1011	Authentication	See section 4.3.1.2
1100	De-authentication	Used by AP / station to end an authentication

Table 4.2: Type 01: Control

Subtype	Name	Explanation
1010	PowerSave-Poll	See section 4.8
1011	Request To Send	See section 4.5
1100	Clear To Send	See section 4.5
1101	ACKnowledgment	See section 4.5
1110	ContentionFree-End	See section 5.4.2 (Related to PCF – defined below)
1111	CF-End + CF-ACK	See section 5.4.2 (Related to PCF – defined below)

Table 4.3: Type 10: Data

Subtype	Name	Explanation
0000	Data	Carries data from higher-layer protocols
0001	Data + CF-ACK	See section 5.4.2 (Related to PCF)
0010	Data + CF-Poll	See section 5.4.2 (Related to PCF)
0011	Data + CF-ACK + CF-poll	See section 5.4.2 (Related to PCF)
0100	Null Data	Frame body carries no data
0101	CF-ACK	See section 5.4.2 (Related to PCF)
0110	CF-Poll	See section 5.4.2 (Related to PCF)
0111	Data + CF-ACK + CF-poll	See section 5.4.2 (Related to PCF)

Table 4.4: Type 11: Reserved

To-DS and From-DS	These two bits together determine whether the frame is intended for the distribution system or a station. Frames transmitted by a station to the AP (and hence the DS) have To-DS set, whereas frames transmitted by an AP (and hence the DS) destined to a station have the From-DS. Note that in an infrastructure BSS, one and only one of the two bits should be set. Both bits set to 0 indicates an IBSS and both bits set to one are used in a wireless bridge environment.
More Fragments	From Figure 4.6, note that the frame body is limited in size to 2312 bytes. If a higher-layer protocol needs to transmit a bigger packet, the 802.11 MAC fragments it. In this case, the initial fragment and any following nonfinal fragments set this bit to one to indicate that there are other fragments following.
Retry	Since wireless is a hostile environment, 802.11 frames may get lost. When this happens, the frame needs to be retransmitted. Retransmitted frames have this bit set to one to help the receiving station detect and eliminate duplicate frames.
Power Management	Since many 802.11 stations are battery operated, 802.11 allows stations to sleep/doze (see section 4.7 for more details). This bit is used by stations to indicate that it is in power-save mode.
More Data	This bit is again related to power management. See section 4.7 for more details.
WEP	802.11 allows frames to be encrypted for security purposes. The protocol used for this is called WEP. For frames encrypted using WEP, this bit is set to one.
Order	The ordering of unicast frames to a station is always maintained by 802.11. However, when both unicast and multicast frames are present, the change of ordering between unicast frames and multicast frames is not guaranteed by default. If the order bit is set to 1, then extra processing in MACs on the transmitting and receiving end can guarantee this too.

4.4.2 Duration/ID

This two-byte field has different meanings depending on the type of frame in which it is used. In most cases, the meaning of this field is "duration" and carries the value of the NAV (network allocation vector), which is used for virtual carrier sensing in the 802.11 MAC. See section 4.5.4. When used to carry the NAV, bit 15 (the most significant bit) in this field is set to 0.

When used in PS-Poll frame, this field contains the Association ID of the association for which the poll is being sent. When used to carry the AID, bit 15 (the most significant bit) in this field is set to 1 and bit 14 is also set to 1.

Finally, when used with point coordination function (PCF) during contention-free periods (CFPs), CFP frames set this field to 32,768 with bit 15 set to 1 and bit 14 set to 0. This allows stations that did not receive beacons announcing the beginning of the contention-free period to use this value as a NAV and delay their transmission so as not to interferre with contention-free transmissions.

4.4.3 Addresses

The 802.11 frame can contain up to four MAC addresses. We first describe the type of MAC addresses possible in 802.11 and then describe where each one of them fits into the 802.11 header. Refer to Table 4.5.

BSSID: Address of the wireless interface of the AP in the BSS.

Destination Address (DA): Address of the final recipient of the frame.

Source Address (SA): Address of the original source of the frame.

Receiver Address (RA): Address of the wireless interface that should process the frame.

Transmitter Address (TA): Address of the wireless interface that transmitted the frame onto the wireless medium. Used only in wireless bridging.

Table 4.5: 802.11 Frame MAC Address Fields

To-DS	From-DS	Address-1	Address-2	Address-3	Address-4
0	0	DA	SA	BSSID	N/A
0	1	DA	BSSID	SA	N/A
1	0	BSSID	SA	DA	N/A
1	1	RA	TA	DA	SA

4.4.4 Sequence Control

The purpose of this field is to help the receiver in discarding duplicate frames and to reassemble fragmented frames. Since it fulfills two functions, the sequence-control field is broken down into a 12-bit sequence number and a 4-bit fragment number (see Figure 4.6).

For the purpose of providing sequence control (and hence duplicate frame detection, etc.), each higher-layer packet is given a sequence number by the 802.11 MAC. It begins at zero and is incremented by one for each higher-layer packet transmitted by the MAC.[4] Since this field is just 12 bits long, it operates as a modulo-4096 counter.

For the purpose of supporting fragmentation (as discussed in section 4.4.1, 802.11 may fragment higher layer packets larger than 2312 bytes). The 4-bit fragment field provides a mechanism for the receiver to reassemble those frames. When a higher-layer packet is fragmented by the 802.11 MAC, all fragments have the same sequence number but each fragment is uniquely identified by a different fragment number. Specifically, the first fragment has a fragment number of zero and each successive fragment has this number incremented by one.

4.4.5 Frame Body

The frame body field is responsible for carrying packets from higher-layer protocols. The frame body can carry packets of up to 2312 bytes or it may be left empty (0 bytes) to generate a Null[5] data packet. Note also that 802.11 makes no assumptions about what higher-layer protocol is running on top of it. In other words, 802.11 can support multiple higher-layer protocols. Now, the 802.11 MAC header itself has no field to carry information about what higher-layer protocol's data it is carrying. This information is needed so that the 802.11 MAC at the receiver knows which higher-layer protocol to hand the data off to. To accomplish this, 802.11 relies on the 802.2 Logical Link Control (LLC) encapsulation. Figure 4.8 shows where the LLC encapsulation fits into the 802.11 frame.

The LLC header is inserted after the MAC header. In LLC Type-1 operation (i.e., unacknowledged connectionless mode), the LLC header is three bytes long and consists of a one-byte destination service access point[6] (DSAP) field, a 1-byte source service access point (SSAP) field, and a 1-byte Control field. The LLC's DSAP and SSAP values of 0xAA indicate that an IEEE 802.2 SNAP header follows. The SNAP Header is five bytes long and consists of a three-byte organizationally unique identifier (OUI) field and a two-byte protocol identifier

---

[4] Since a MAC layer transmission is still handling the same higher-layer packet, a retransmitted packet must use the same sequence number

[5] We will see why this packet may be useful in later chapters.

[6] Note that this "access point" terminology has nothing to do with the 802.11 access point. Instead, it refers to the points where higher layers access the LLC for access. See Figure 4.1.

DSAP (1 byte)	SSAP (1 byte)	CTRL (1 byte)	OUI (3 bytes)	PID (2 bytes)

Octets: 2	2	6	6	6	2	6	0-2312	4
Frame Control	Duration/ ID	Address 1	Address 2	Address 3	Sequence Control	Address 4	Frame Body	FCS

MAC Header

Fragment Number	Sequence Number

Bits   4   12

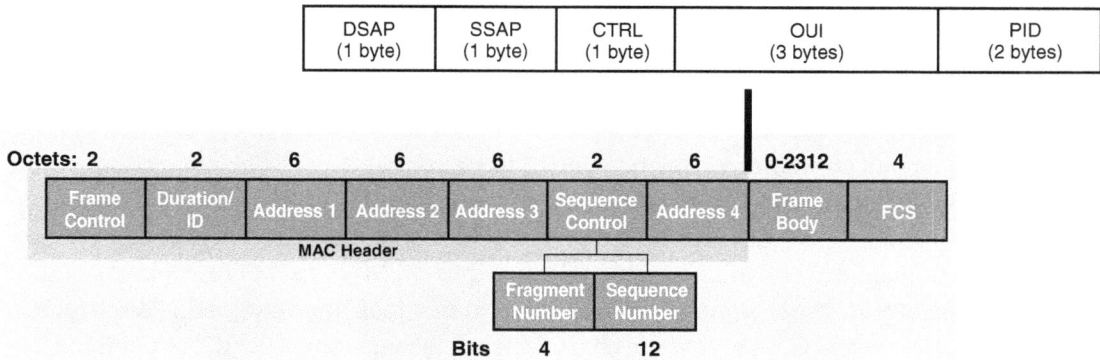

Figure 4.8: 802.11 Logical Link Control Header

(PID). An example will make things more clear. For carrying IP/ARP datagrams over 802.11, the value of these fields would be:

DSAP and SSAP: 0xAA → SNAP header follows

CTRL: 0x3 → Unnumbered information (since IP is a best-effort delivery service)

OUI: 0x00-00-00 → The PID is an EtherType—i.e., a routed non-OSI protocol.

PID: 0x08-00 → IP datagram is carried in this frame.

PID: 0x08-06 → ARP packet is carried in this frame.

4.4.6 Frame Check Sequence (FCS)

The FCS field is also known as CRC (cyclic redundancy check) because the CRC algorithm is used to calculate a checksum of all fields in the MAC frame. This checksum is calculated by the source and appended to the frame. On receiving a frame, the receiving station also calculates the CRC of the frame and compares it with the attached CRC. This allows the receiver to verify that the packet has not been corrupted while in transmission.

4.5 Accessing the Medium

Wireless is a shared medium. This means that all stations wishing to access the wireless medium should follow a protocol to decide which station would transmit when multiple stations compete for the medium. The protocols which define these rules are known as media access protocols.

The 802.11 standard defines two media access protocols: DCF and PCF. Distributed co-ordination function (DCF), as its name suggests, allows multiple stations to interact (and coordinate) access to the medium without a central control. Note that, in an infrastructure BSS, this means that even though all communication must pass through the "central" AP, the AP does not have any special status as far as access to the medium is concerned. It must coordinate and/or compete with all other stations in the BSS to gain access to the channel. On the other hand, the point coordination function (PCF) lets a central entity—the point coordinator (or PC, usually located in the AP)—control the medium. The PC is responsible for managing access to the medium. All 802.11-compliant devices must support DCF, whereas the support for PCF is optional. DCF is by far the most widely deployed MAC in 802.11 networks and is discussed below. PCF was intended for real-time applications like voice and is discussed in Chapter 5. However, PCF has never been, in fact, implemented by any 802.11 equipment manufacturers.

4.5.1 CSMA-CD

To understand the DCF, it is easier to start with the IEEE 802.3 MAC (wired Ethernet) as a starting point. The 802.3 protocol specifies a CSMA-CD (carrier sense multiple access with collision detection) media access protocol. When a station wants to transmit on the medium, it first uses CSMA to detect whether the network is idle. It does this by "sensing" the channel (i.e., measuring the voltage level) for ongoing transmissions. A station transmits only if CSMA concludes that no other station is currently transmitting. To resolve the case where multiple stations may simultaneously sense the channel as idle, collision detection is used. To use CD, the station continues to "sense" the channel after it starts transmitting. If another station also starts transmitting, this would result in a "collision" (i.e., garbled voltage levels on the wire). Since both stations are sensing the channel for voltage fluctuations, the collision would be detected by both stations. Both stations would now stop transmission and wait for a random time (this is called backing off) before trying to access the medium again. The amount of time each station backs off is based on an exponential back-off algorithm.

4.5.2 Wireless Media Access Challenges

The CSMA-CD scheme works well for a wired medium. However, this approach is unsuitable for the wireless medium for multiple reasons.

1. Implementing a collision detection would require the implementation of a full-duplex radio (capable of transmitting and receiving at the same time). This would increase the cost of the equipment considerably and is therefore not considered an option.

2. Two, the wireless medium is inherently open to interference from a wide variety of sources, especially since 802.11 operates in the unlicensed frequency spectrum and differentiating between collisions and interference is not a trivial task.

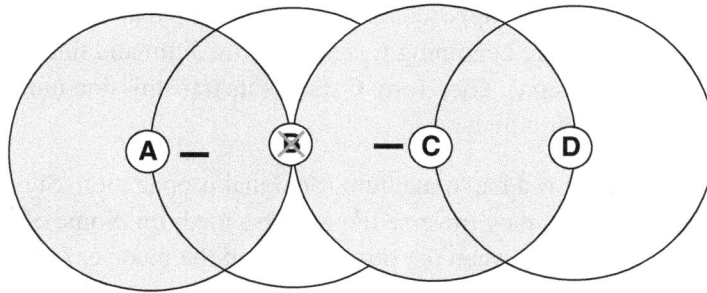

Figure 4.9: Hidden Terminal Problem

3. In a wireless medium, we cannot assume that all stations can hear each other. Therefore, just because a station willing to transmit senses the medium as free does not mean that the medium is free around the receiver area. Figure 4.9 shows three wireless terminals: A, B and C. The radio transmission range of each terminal is shown by a circle around the terminal. As is clear, terminal B lies within the radio-transmission range of both terminals A and C. Consider now what happens if both A and C want to communicate with B. CSMA requires that, before starting transmission, a terminal "senses" the medium to ensure that the medium is idle and therefore available for transmission. In our case, assume that A is already transmitting data to B. Now, C also wishes to send data to B. Before beginning transmission, it senses the medium and finds it idle since it is beyond the transmission range of A. It therefore begins transmission to B, thus leading to collision with A's transmission when the signals reach B. This problem is known as the hidden terminal problem since, in effect, A and C are hidden from each other in terms of radio-detection range.

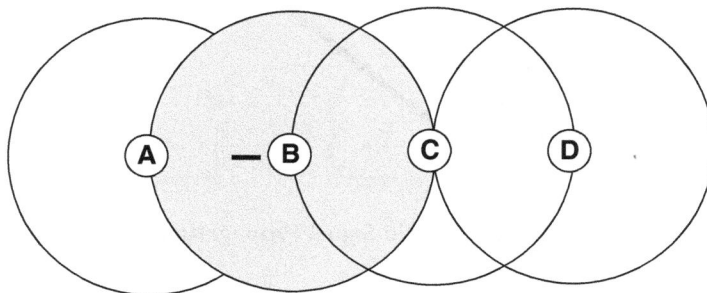

Figure 4.10: Exposed Terminal Problem

4. Consider what happens when B wants to send data to A and C wants to send data to D (Figure 4.10). As is obvious, both communications can go on simultaneously since they do not interfere with each other. However, the carrier-sensing mechanism raises

a false alarm in this case. Suppose B is already sending data to A. If C wishes to start sending data to D, before beginning it senses the medium and finds it busy (due to B's ongoing transmission). Therefore, C delays its transmission unnecessarily. This is the exposed terminal problem.

5. The wireless medium is a harsh medium for signal propagation. Signals undergo a variety of alterations as they traverse the wireless medium. Some of these changes are due to the distance between the transmitter and the receiver (attenuation), others are due to the physical environment of the propagation path (slow fading, interference) and yet others are due to the relative movement between the transmitter and the receiver (multipath effects, fast fading). All these adversities result in a very high (typically 5–15%) packet-error rate in the wireless medium.

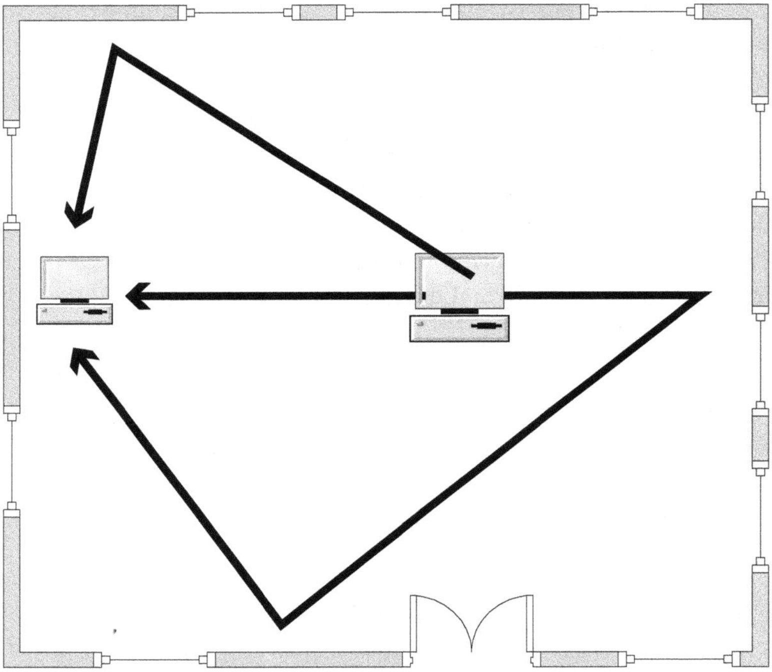

Figure 4.11: 802.11 Radio Signal Propagation Examples

To solve these issues, 802.11 uses several mechanisms which we discuss in the next few subsections.

4.5.3 Positive ACK

To overcome the issues of operating in a high packet-error-rate environment, 802.11 relies on a positive acknowledgment[7] scheme, where each unicast data transmission must be explicitly acknowledged by the receiver. The onus of error recovery thus falls on the transmitter—it must ensure that each unicast data frame that it transmits is acknowledged (aka ACKed) by the receiver. If the transmitter does not receive an ACK, it must assume that the packet is lost and must retransmit.

4.5.4 NAV

As discussed, carrier sensing in the wireless medium is complicated by the presence of various sources of interference and scenarios like hidden node and exposed node. To improve carrier sensing, 802.11 employs a virtual carrier-sensing mechanism. This is achieved using the Duration field in the MAC header (see section 4.4.2). The value in the duration field represents the number of microseconds that the medium is expected to remain busy. Other stations that "hear" this frame consider the medium to be busy for this duration.

Note that the virtual carrier-sensing mechanism is used along with (and NOT instead of) the physical carrier-sensing mechanism to get a better state of the medium. If either the physical or the virtual carrier-sensing function indicates that the medium is busy, the medium is considered busy.

4.5.5 CSMA-CA

The CSMA-CA algorithm forms the backbone of the DCF. The CSMA-CA (carrier sense multiple access with collision avoidance) algorithm replaces the collision detection mechanism with collision avoidance. Collision avoidance is achieved as follows:

a. If the station senses[8] that the medium has been longer than a period of time referred to as the DCS Inter-frame Spacing (DIFS),[9] it can begin transmitting frames immediately.

b. If the station senses that the medium is busy, the station must wait for the channel to become idle. Moreover, when the station does sense that the channel is now idle, it does not start transmitting immediately (as in CSMA-CD); instead the station keeps sensing the channel for an additional random time before it actually starts

---

[7] Compare this with negative acknowledgment schemes where, by default, transmitted packets are not acknowledged by the receiver. Instead, only when the receiver detects packet loss does it trigger a retransmission.

[8] This includes both physical and virtual carrier sensing.

[9] Note that this would typically mean that this station is the only station wishing to transmit at this moment in the BSS.

transmitting. This is the collision avoidance part of the algorithm: since each station waits a random time after sensing the channel idle, the probability of collision decreases. An example will make things clearer.

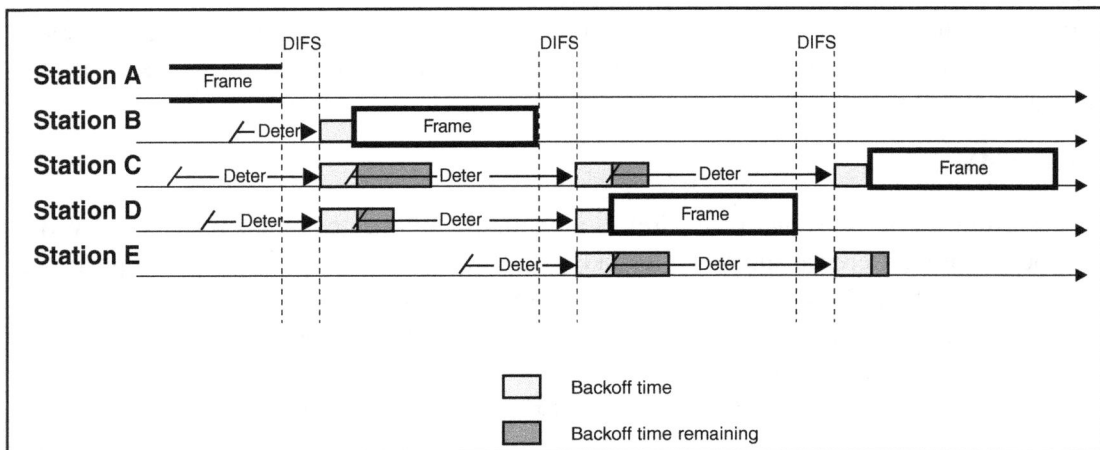

Figure 4.12: 802.11 CSMA-CA Example

From Figure 4.12, note the following salient features of CSMA-CA:

1. While A is transmitting, B, C, and D get data to transmit but they defer since channel is occupied by A.

2. After sensing the channel idle for the first time, all stations wait for a minimum period of DIFS (DCF inter-frame spacing).

3. After the DIFS expires, each station keeps sensing the channel for an additional random time. The duration of this random time is a multiple of a slot time (the fundamental time unit in the 802.11 PHY, equal to the minimum amount of time any station can detect a packet transmission from any other station).

4. The number of additional slot times that a station should wait additionally (after detecting the channel idle) is determined by a BC (back-off counter).

5. The value BC is randomly selected from the range [0, CW], where CW refers to the contention window.

6. The value of CW is doubled after each unsuccessful (un-ACKed) transmission, thus reducing the probability of collision when multiple stations are attempting to access the channel simultaneously.

7. The minimum and maximum values of CW (CW$_{min}$ and CW$_{max}$ respectively) are specified by the PHY standard in use. For example, CW$_{min}$= 21 and CW$_{max}$ = 1023 for 802.11b.

8. On the other hand, after each successful (ACKed) transmission, the station should again wait CW (again selected randomly) slot times before starting the next transmission. This prevents system hogging by a station once it gets access to the channel.

9. In Figure 4.12, this random time happens to be smallest for station B and therefore B gets to transmit next.

10. Once B finishes, C, D and E now have data to transmit. Again, after sensing the channel idle for the first time, all stations wait for a minimum period of DIFS.

11. However, stations that had deferred their transmission last time (C and D) continue to count down from their previous value instead of starting a new countdown. This ensures that C and D get a higher priority than E, who just started competing.

4.5.6 Inter-Frame Spacing (IFS)

In the last section we said that each station that wishes to transmit waits at least till it senses the channel idle for DIFS (DCF IFS) time interval. This is not completely true. The 802.11 standard defines four different IFS periods. Varying IFS creates different priority levels for different types of traffic, since higher-priority traffic can be made to wait for a smaller time (SIFS) than lower-priority traffic (DIFS) for access to the medium as shown in Figure 4.13.

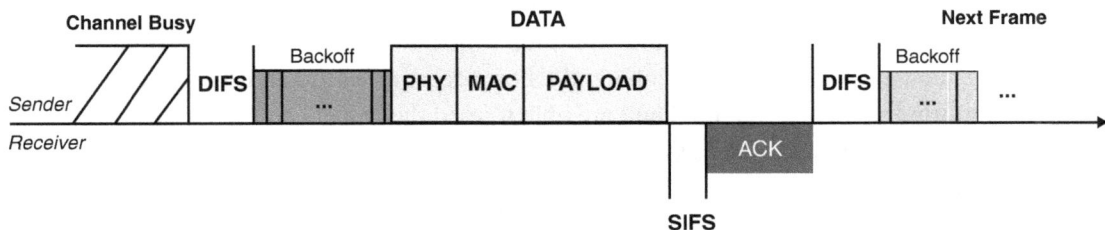

Figure 4.13: Various Inter-Frame-Spacings in 802.11

802.11 specifies four IFSs, which can be used depending on the context: SIFS, DIFS, PIFS and EIFS with SIFS < PIFS < DIFS < EIFS.

SIFS (Short IFS): Used for the highest-priority transmissions. These higher-priority transmissions are ACKs (see section 4.5.3), RTS-CTS (see section 4.5.7) and fragments of a fragmented packet. Therefore, the ACK of a transmitted frame is considered a higher priority than the transmission of another data frame.

PIFS (PCF IFS): Used by PCF during contention-free operation. See section 5.4.2.

DIFS (DCF IFS): The default IFS used in DCF for resolving contetion of frames to the wireless medium. See section 4.5.5.

EIFS (Extended IFS): This IFS is used in error scenarios. If a station receives a packet from which it cannot determine the value of the NAV/Duration-field, it must defer transmission for EIFS to avoid collisions with future packets belonging to the same dialog.

4.5.7 RTS-CTS

802.11 specifies an RTS-CTS mechanism that can be used along with NAV to solve the hidden-node and exposed-node problems as shown in Figure 4.14.

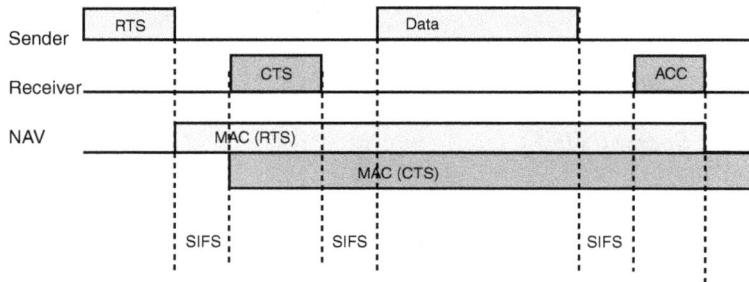

Figure 4.14: 802.11 RTS/CTS Example

In this approach, when a station has data to transmit and gets access to the medium after the CSMA-CA procedure, it sends an *RTS (Request To Send)* frame instead of the data itself. The RTS frame contains the duration (i.e., the NAV) for which the transmitting node wants to capture the channel. Since wireless is inherently a broadcast medium, all nodes in the vicinity of the transmitting node hear the RTS and are able to read the NAV. Therefore, these nodes consider the channel busy for the NAV mentioned. Now, when the RTS reaches the receiver, it responds with a CTS (Clear to Send) frame which contains a NAV value too. All nodes in the vicinity of the receiver hear the CTS and accordingly assume the channel busy for this duration. There are two important things to note here. First, the NAV values in the RTS and CTS frames include the time that the medium would be used for transmitting the data packet and the corresponding acknowledgment.

To see how this mechanism solves the hidden-node problem, recall the problem from Figure 4.9. If the nodes A, B and C used the RTS/CTS protocol, A would transmit an RTS before beginning its transmission. C would not hear this RTS since it's outside the transmission range of A. However, when B responds with a CTS, this CTS would reach C. Node C can therefore delay its transmission to avoid collision at B. The duration of the delay is obtained from the NAV in the RTS/CTS messages.

Note that RTS/CTS messages are not a required feature in 802.11 MAC and a node can directly send data onto a channel. The decision of whether or not to use RTS-CTS is a factor of the packet sizes being used and the load and capacity in the system. The advantages of RTS-CTS can be realized for large packets[10] and come at the cost of additional frames and thus reduced system capacity.

4.6 802.11 PHY

One of the original design decisions of 802.11 was to have a clear interface between the MAC and PHY layers so that multiple PHY layers could be used with the same MAC. In retrospect, this has been one of the reasons for the success of 802.11. To deal with multiple physical layers, all of which use the single 802.11 MAC layer, the 802.11 standard splits the physical layer into two sublayers: the PLCP and the PMD. The *Physical Layer Convergence Protocol (PLCP)* is responsible for appending the PHY layer header and providing a uniform interface of the various physical layers (and the PMDs) to the 802.11 MAC layer. The *Physical Medium Dependent (PMD)* is responsible for actually transmitting the frames onto the wireless medium.

The various PHY layers differ in the choice of the modulation scheme and the choice of the spread-spectrum technique. Modulation refers to the conversion of an 802.11 frame from the digital to analog domain. The 802.11 frame discussed until now was described in terms of bit formats, since this is how the higher layers see it. However, what needs to be transmitted over the medium is analog signals. Different PHY layers use different modulation schemes to achieve this. There is a wide variation in these choices: PSK (phase shift keying), FSK (frequency shift keying), CCK (complementary code keying), PBCC (packet binary convolutional coding), QAM (quadrature amplitude modulation).

The other interesting difference between the various PHYs is the choice of spread-spectrum technique. Spread spectrum is the technique of diffusing the signal power of a narrowband signal over a large range of frequencies at the transmitter and performing the inverse operation at the receiver. This "spreading" of the signal makes the transmission more resistant to interference from noise and other signals. There are three basic spectrum techniques used in the various 802.11 PHYs: FHSS (frequency hopping spread spectrum), DSSS (direct sequence spread spectrum) and OFDM (orthogonal frequency division multiplexing).

The original base 802.11 standard ratified in 1997 supported three separate PHY layers that provided for data rates of 1 to 2 Mbps. However, as the technology and standards have evolved, several new PHY layers have been proposed and ratified. As of the writing of this

---

[10] Using the RTS-CTS mechanism to reserve the medium for packets that are the same size as RTS/CTS frames themselves obviously does not make sense.

book, 54-Mbps PHY-based products are available in the market and the 802.11n standard (which promises hundreds of megabits per second) is in the final stages of ratification.

Any attempt to discuss the various PHY layers in any justifiable detail would require a separate book, and the scope of this book does not justify such a discussion. Therefore, we do not discuss the various PHY layers in depth. Instead, we look at those aspects of the PHY layer that are relevant to the MAC layer and layers above that. This is important since the choice of a PHY layer has several ramifications on the capacity and capabilities of an 802.11 deployment. The three most widely deployed PHY layers today are 802.11b, g and a. We therefore limit our discussion to these three PHYs. Table 4.6 shows a high-level comparison of the common 802.11 PHYs and the following subsections discuss some important aspects from the table.

Table 4.6: 802.11 Physical Layer Characteristics

	802.11b	802.11g-only	802.11a
Operating Frequency (GHz)	2.4	2.4	5.0
Supported Data Rates (Mbps)	1,2,5.5,11	6,9,12,18,24,36,48,54	6,9,12,18,24,36,48,54
Modulation	PSK / CCK		PSK / QAM
Spread Spectrum	DSSS	OFDM	OFDM
Basic Rate (Mbps)	2	6	6
Channels			
Preamble (µs)	144		20
PLCP (µs)			4
SIFS (µs)	10	10	16
Slot Time (µs)	20	9	9
CWmin	32	16	16

Note:
PIFS = SIFS + Slot_Time
DIFS = PIFS + Slot_Time

4.6.1 PLCP Framing

Section 4.4 discussed 802.11 framing in detail. However, we covered only the MAC (and LLC) header in that discussion. The PHY layer (specifically the PLC) adds its own headers before transmitting a frame. In this section, we discuss the PLCP framing used in various 802.11 PHYs.

4.6.1.1 802.11b

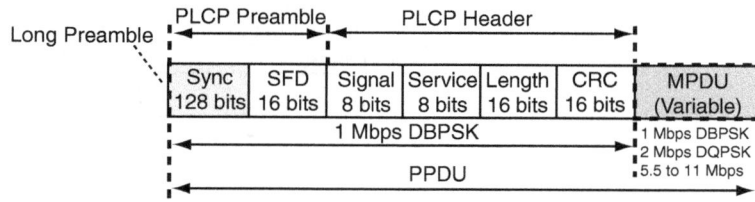

Figure 4-15: 802.11 PLCP Header with Long Preamble

Figure 4.15 shows an 802.11b PHY frame. The MPDU shown in the figure is where the 802.11 MAC frame goes. The PHY header consists of two parts, a PLCP preamble and a PLCP header. The PLCP preamble allows the transmitter and the receiver to synchronize and derive common timing relationships. It is composed of the Sync field, which is composed entirely of 1s, and the SFD (Start Frame Delimiter) field, which allows the receiver to find the start of the frame, even if some of the sync bits were lost in transit. This field is set to 0000 0101 1100 1111.

The PLCP header consists of four fields. The Signal field indicates the transmission rate used for this frame. As explained in section 4.6.2, each PHY supports multiple transmission rates, thus allowing each frame to be transmitted at a different rate. The Service field is reserved for future use. The length field indicates the number of microseconds required to transmit the frame. Finally, the CRC field protects the header frame against corruption by carrying a 16-bit CRC of the contents of the four header fields.

Note from Figure 4.15 that different parts of an 802.11 frame are transmitted at different rates. For example, in Figure 4.15 the PHY header is transmitted at 1 Mbps, whereas the MPDU may be transmitted at any of the rates supported by the 802.11b PHY (1, 2, 5.5. or 11 Mbps). This is an important observation which we will discuss in later chapters.

802.11b supports two types of PHY framing: long preamble, as shown in Figure 4.15, and the short preamble, shown in Figure 4.16. The short-preamble frame contains the same fields as the long-preamble frame but with some important differences. First, the sync field is smaller in size (56 bits instead of 128 bits). Second, the PLCP header is transmitted at 2 Mbps instead of 1 Mbps. Third, the MPDU can be transmitted at 2, 5.5 or 11 Mbps. These faster transmission rates and smaller headers make short preamble the more efficient in terms of bandwidth utilization.

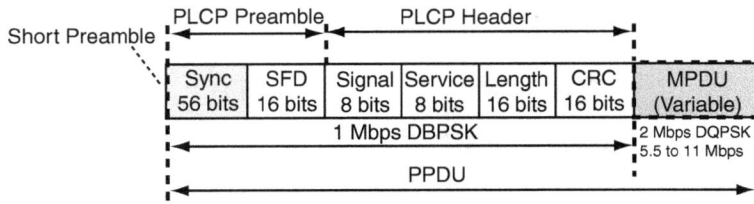

Figure 4.16: 802.11 PLCP Header with Short Preamble

4.6.1.2 802.11g/a

Figure 4.17 shows an 802.11g/a PHY frame. The PSDU field shown in the figure is where the 802.11 MAC frame fits in. The PHY header consists of four parts—a PLCP preamble, a PLCP header, a Tail and a Pad.

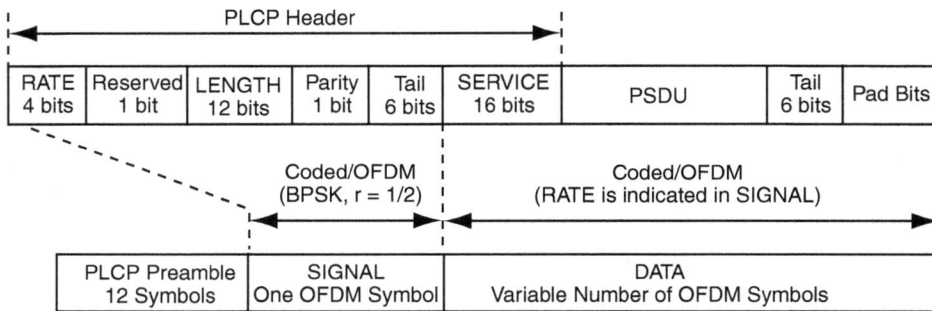

Figure 4.17: 802.11 a/g PLCP Header

The PLCP preamble, as in 802.11b, allows the transmitter and the receiver to synchronize and derive common timing relationship.

The PLCP header consists of two fields: Signal and Service. The Signal field is further broken down into five subfields: Rate indicates the transmission rate used for this frame; Length indicates the number of bytes in the embedded MAC frame; Parity is an even parity bit for the first 16 Signal bits to protect against transmission errors; Tail bits are all set to 0 and the Reserved bit is reserved and set to 0. The Service field is set to 0 and is used for initializing the PHY scrambler.

The Tail field in the Trailer is needed for smooth convolution coding and the Pad field is needed since OFDM requires that, given the data rate of transmission, the size of the data block be fixed.

4.6.2 Transmission Rate

The most "advertised" difference in the various PHY standards proposed by 802.11 is the raw transmission rate, which in turn affects the application layer throughput. 802.11b offers a maximum transmission rate of 11 Mbps whereas 802.11g and 802.11a offer a maximum transmission rate of 54 Mbps. However, note that each PHY layer standard supports a range of transmission rates. For example, 802.11b supports transmission rates of 1, 2, 5.5 and 11 Mbps and 802.11a supports transmission rates of 6, 9, 12, 18, 24, 36, 48 and 54 Mbps. To understand the motivation behind this, realize that the different transmission rates within a PHY standard are achieved by using various modulation schemes.

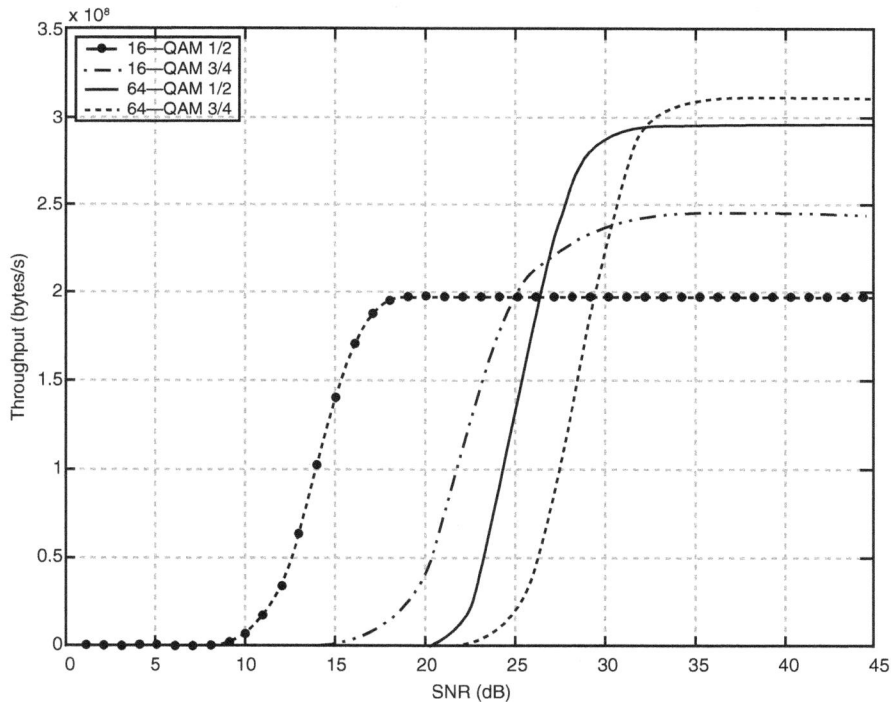

**Figure 4.18: Throughput vs. Signal-to-Noise Ratio (SNR)
for some 802.11a Modulation Schemes[11]**

---

[11] Automatic IEEE 802.11 rate control for streaming applications, Haratcherev et al., *Wireless Communication & Mobile Computing*, 2005.

More complex modulation schemes that achieve higher transmission rates do so by using "aggressive" modulation techniques. For example, 802.11a uses 64-QAM 3/4 for achieving a transmission rate of 54 Mbps. However, these aggressive modulation techniques also have a higher sensitivity to channel conditions. They require a high SNR (signal-to-noise ratio) to achieve these high data rates. Such high SNR channel conditions are not always possible in the wireless medium. Therefore, each PHY also specifies "conservative" modulation techniques which achieve lower data rates but can operate under lower SNR. For example, 802.11a uses BPSK 1/2 for achieving 6 Mbps. Since the SNR decreases as the distance between the communicating stations increases, another way to look at this is to say that supporting multiple transmission rates allows for a coverage (SNR) versus capacity (data rates) trade-off.

Hence, as the SNR decreases, it is better to use lower transmission rates with modulation schemes that are less sensitive to channel noise. This concept of dynamically adjusting the transmission rate based on channel conditions and performance is known as *rate adaptation*. The algorithm to be used for the selection of the "appropriate" transmission rate is not specified in the 802.11 standards and is expected to be one of the product differentiators among various 802.11 vendors.

4.6.3 Nonoverlapping Channels

The choice of the PHY layer also determines the number of nonoverlapping channels available in a WLAN. This, in turn, affects the system capacity of a WLAN. The concept of nonoverlapping channels (based on frequency-reuse) is not new to wireless networking. Cellular networks have exploited this concept to support an increasing subscriber density. The frequency reuse concept is based on the fact that signals attenuate as they travel through the wireless medium. This allows the same frequency (i.e., channel number) to be reused provided the geographical area (cells) where the same frequency is being used are far enough apart not to cause interference[12] with each other. Thus, the minimum distance between two cells where the same channel number can be used is determined by the number of nonoverlapping channels. This is true in both cellular networks and in WLANs. For example, instead of all users (stations) in a given geographical area (cell) having only one access point to associate with, if a second access point is installed in the "same area" (i.e., within the radio range of the first cell) but on a nonoverlapping channel, those users (stations) will now be shared among the two access points, effectively doubling the system capacity. However, if access points are installed within radio range of each other and are not configured on nonoverlapping channels, then the overall system performance is actually reduced due to increased contention for the channel (and therefore increased back-off times) and increased noise floor (and therefore increased retransmissions)—both of which waste bandwidth.

---

[12] This sort of interferrence is known as cochannel interference.

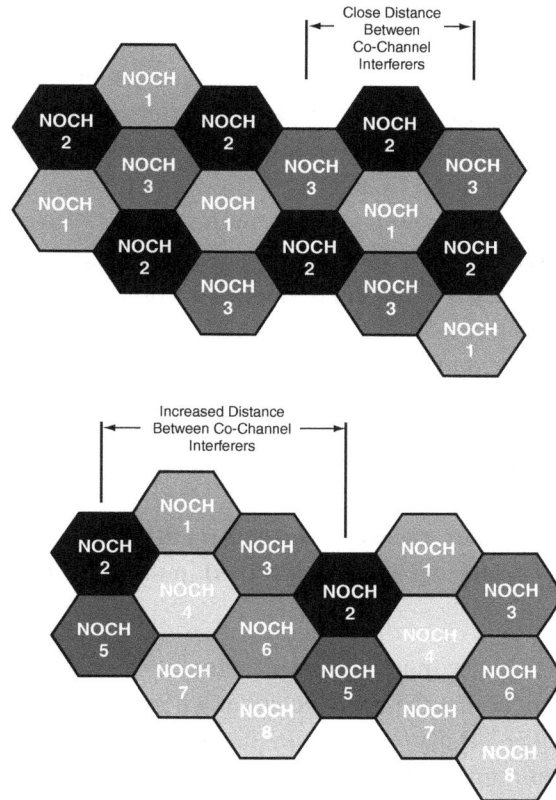

Figure 4.19: Example 802.11 Radio Cell Layout (top: un-optimized, bottom: optimized)

The number of nonoverlapping channels available for a certain 802.11 PHY is determined by regulatory considerations. For example, in 802.11b and 802.11g the number of nonoverlapping channels available is three. This is due to the spectrum-availability limitations imposed by their operation in the 2.4-GHz frequency band. Radio-frequency regulations governing the use of wireless LANs in the 2.4-GHz frequency band restrict systems operating within this band to three (3) nonoverlapping channels. On the other hand, 802.11a, operating in the 5-GHz frequency band, allows for at least eight[13] nonoverlapping channels. Thus, 802.11 networks with higher capacity (i.e., user density) requirements would be better off with using 802.11a as the PHY layer.

---

[13] The number might vary from country to country.

4.6.4 Power Consumption

The choice of 802.11 PHY also affects the power consumption of an 802.11 device. Given that a large number of 802.11 devices are expected to be battery-operated, this becomes an important consideration. At first, it may seem instinctive to conclude that PHYs using complex modulation schemes (like 802.11g and 802.11a) would be more power intensive due to the complex processing that they may require. Even though this observation is correct at its face value, taking a big-picture view changes things. Realize that although requiring more power per bit, actual transmission and reception take less time for a given packet size, thanks to the higher 802.11g/a data rates. Thus, the 802.11 RF transmitter will spend less time in the awake state and can be powered off faster. We will discuss power consumption in the next section and in detail in Chapter 9.

4.7 Power Save in 802.11

The 802.11 specification provides a power-management algorithm for stations to conserve power. The process of power management as specified in the 802.11 standard is shown in Figure 4.20.

Even though Figure 4.20 is pretty much self-explanatory, there are some important aspects worth pointing out. First, note that the decision about when a station decides to doze is left entirely up to the station: the AP plays no part in this decision. The station is only required to inform the AP of its power mode and it does this by setting the Power Management bit in the MAC header (see section 4.4.1) of the last frame that it sends to the AP. Note that this bit is overriding in that the value of this bit in Frame B overrides the value in Frame A, if Frame B is sent after Frame A. Therefore, if a station in power-save mode (aka a dozing station) wakes up to transmit 802.11 frames, it must set the Power Management bit in the MAC header of each of these frames if it continues to stay in doze mode. If any of the frames has this bit cleared, the AP will interpret it to mean that the station has exited its doze mode.

Note that Figure 4.20 shows that the station must wake up on every beacon to interpret the TIM field in the beacon that tells it whether or not the AP has packets buffered for it. This is not strictly necessary. 802.11 allows the station to wake up every N^{th} beacon (with $N > 1$). This N is known as the listen interval and its value must be negotiated between the station and the AP at association time. The Association Request frame from the station to the AP carries a Listen Interval field that the station requests. Realize that larger values of N mean more buffer space at the AP, since it requires APs to potentially buffer frames destined for stations while the station is dozing. The trade-off here is that larger Listen Intervals allows for better power savings and longer battery life at the cost of longer packet latency and reduced throughput.

Finally, note that when the station realizes that the AP has packets buffered for it, the station uses the PS-POLL control frames (see section 4.4.1) to retrieve these buffered packets from

STA   **Access Point**

STA decides to Doze
when it is idle   Beacon

Data / Null-Data Frame;
STA goes into Doze Mode.   Power-Save bit = 1 in MAC header.

AP is now responsibile for buffering
any packets destined for this STA.

Beacon

Beacon

Dozing STA wake up
periodically to receive
Beacons

Beacon

AP receives packets destined for
this STA and buffers them.

AP uses the TIM / DTIM field in the Beacon to
inform the STAs about packets it has buffered   Beacon

STA reads the TIM / DTIM
and determines that the   STA uses PS-POLL to
AP has packets buffered   retrieve buffered packet(s).
for it.

Beacon

Data Frame;
More bit = 1 in MAC header

STA uses PS-POLL to
retrieve buffered packet(s).

Beacon

Data Frame;
More bit = 1 in MAC header

STA uses PS-POLL to
retrieve buffered packet(s).

End Of Buffered Packets at AP

Data Frame;
More bit = 0 in MAC header

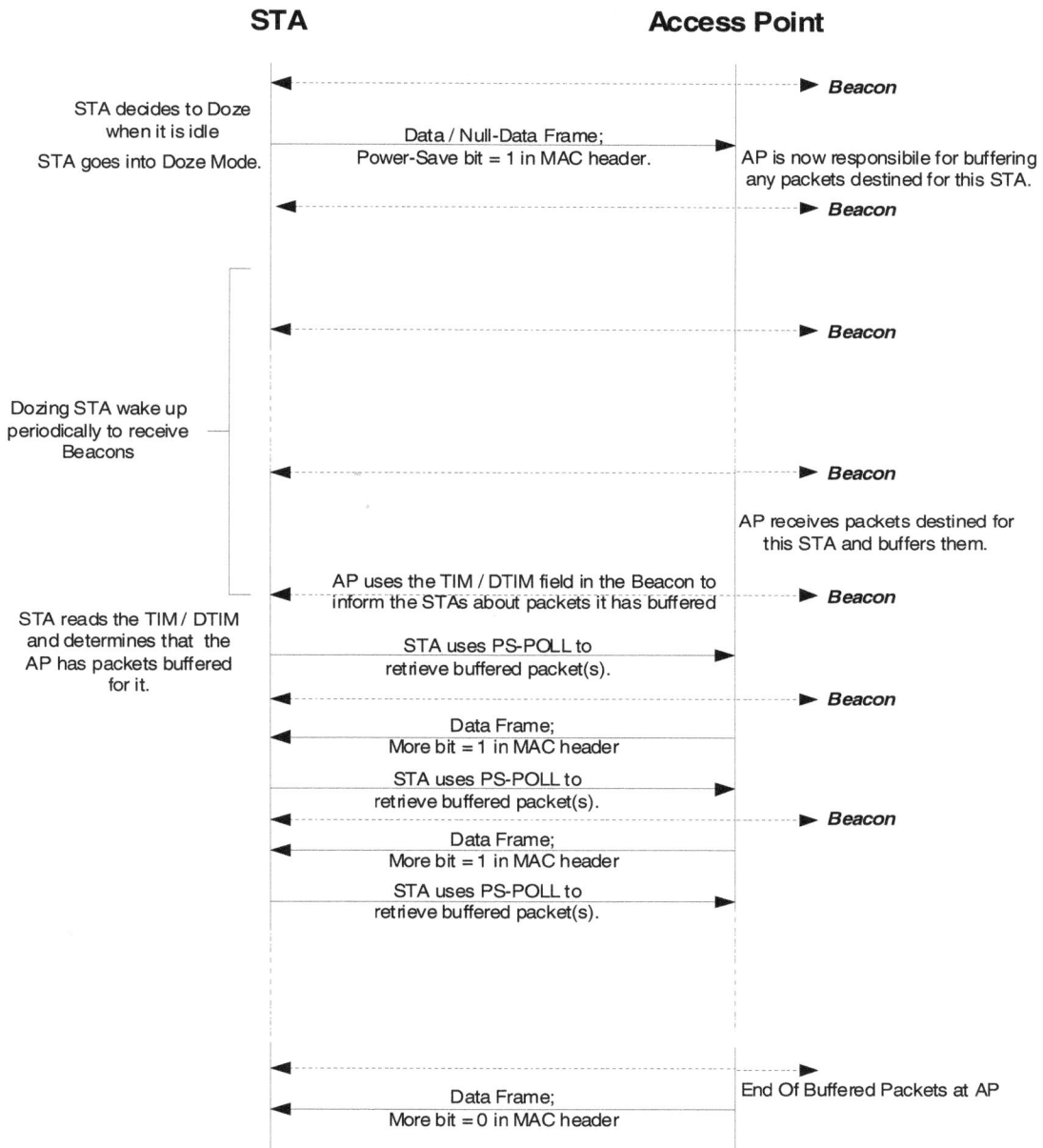

Figure 4.20: 802.11 Power Save Operations

the AP. After receiving data in response to these polls, a powersave STA has two choices. It may either continue to be in doze mode or it may decide to exit its doze mode. The former approach is suitable for low-intensity (packets-per-second) traffic with high delay tolerance, whereas the latter approach is suitable for high-intensity traffic with low delay tolerance.

4.8 Conclusion

This finishes our discussion of 802.11 and WLANs. The next chapter looks at the challenges of "putting" VoIP over WLANs and the rest of the book will discuss solutions to these challenges.

VoWLAN Challenges

5.1 Introduction

As discussed in Chapter 4, the 802.11 standard has come to be somewhat of an "umbrella protocol" due to the numerous enhancements (e, g, h, k, etc.) that have been ratified (or are being discussed) to improve the base 802.11 standard, which was ratified in 1997. Given this situation, it is very difficult to determine what kind of 802.11 deployment is being discussed when someone uses the generic term of WLAN or VoWLAN. The approach taken in this book is to discuss the challenges of "Voice-over-802.11b" networks in this chapter and then discuss possible solutions (including 802.11 enhancements) to these challenges in subsequent chapters. The reasoning behind this approach is simply that 802.11b is by far the most widely deployed WLAN standard today.

5.2 VoWLAN

As we saw in Chapter 3, VoIP comes in many flavors. At a high level, we can distinguish between these flavors by considering where in the overall architecture voice transitions from the PSTN to the IP network. At one extreme is the traditional PSTN model, where one black phone calls another black one and the voice call is established entirely using the PSTN. There is no IP and hence no VoIP in this scenario. At the other end of the extreme is the end-to-end VoIP model where an IP "phone" (which may be a soft-phone application running on a PC or an actual physical entity) calls another IP phone without the voice call ever transitioning to the PSTN. Somewhere in the middle of these two extremes lies the concept of gateways, which connect the IP world to the PSTN.

Note how the discussion of VoIP is restricted to the wired domain. Sure, it is possible to install a media gateway that carries calls from/to wireless cellular subscribers over an IP network, and it is also possible to use cordless phone technology in VoIP architectures, but the VoIP end-device itself is restricted to being a wired device. In other words, IP phones (soft or hard) are always wired devices. This limited deployment of VoIP in "mobile" scenarios. This is where VoWLAN comes in. The wide-scale deployment of 802.11 networks means that it is now possible to implement a VoIP solution over a WLAN instead of a wired LAN like Ethernet and VoIP can, for the first time become a wireless solution.

At the face of it, the solution of running VoIP implementations over 802.11 instead of 802.3 (Ethernet) seems like a simple proposition. After all, one of the primary design criterion for the OSI-layered architecture was to minimize the interdependence between layers. Arguably, since VoIP is implemented at Layer 3 and above, a change in the layer-2 protocol should be trivial. However, this is far from the case. At the outset, let us realize that the 802.11 standard was designed primarily for data communication. However, voice communication is inherently very different from data communication. Unlike data, voice traffic is characterized by small packets transmitted periodically and symmetrically in both directions. Voice traffic also has its own constraints in that it is extremely sensitive to delay and jitter. Furthermore, the quality of a voice call is also dependent on the packet-loss characteristics: while small losses can be tolerated, large gaps (bursty packet losses) will cause serious degradation in voice quality. To use 802.11 for voice communication therefore poses some major challenges. This chapter discusses in detail the challenges of integrating VoIP with 802.11.

5.3 System Capacity and QoS

This section deals with system capacity and quality of service (QoS) issues in VoWLAN. We start with system capacity. Defining system capacity is a tricky issue. It is often described in terms of channel bandwidth (Mbps). However, this can be a misleading parameter. We see why this is so in section 5.3.1. Here, we just want to emphasize that, for VoWLAN systems, the simplest and most useful definition of system capacity would simply be the number of simultaneous voice calls that can exist in a BSS. This is the definition we use.

Even though it may not be clear at first why the topics of system capacity and QoS are clubbed together, a little analysis will reveal that these two topics are inherently linked together. In VoIP, the term QoS is usually used to refer to the real-time requirements (low delay, low jitter and loss-characteristics, etc.) of voice, video and so forth. The basic approach to achieving QoS is to "mark" real-time packets so they get prioritized access to network resources like bandwidth. This may or may not involve reserving resources in the network for real-time traffic. However, the basic philosophy is that, since network resources are limited, real-time traffic should have prioritized access to it. Note that if there are "enough" network resources available for all traffic, there is no need to prioritize real-time traffic. Hence, the concept of system capacity and QoS are inherently linked. If we have enough system capacity, there is no need for QoS mechanisms. This is, for example, the case when making VoIP calls within a LAN[1] that uses 100 Mbps or Gigabit Ethernet. This is one of the reasons why QoS has traditionally been a Layer-3 (or above) issue in VoIP.

---

[1] LAN here means strictly LAN. I.e., VLANs are not included in this definition. In fact, VLANs are one of the primary users of Layer-2 QoS mechanisms.

Another reason for treating QoS in the higher layers is because most VoIP implementations simply treat the IP network as a "cloud" without any information about the underlying link layer, since the VoIP endpoints do not know about what happens (for example, what Layer-2 technology is used) in the cloud. This is not to say that VoIP implementations never use Layer-2 QoS. There are scenarios where VoIP endpoints are aware of the Layer-2 technology being used and Layer-2 bandwidth is at a premium. In such scenarios (VoCable, for example) Layer-2 QoS (DOCSIS in VoCable) has been used in VoIP deployments. Since in VoIP over WLAN we also know the characteristic of the underlying link layer, QoS becomes relevant at Layer 2. The following subsections discuss why system capacity and QoS are important issues in VoWLAN.

5.3.1 Packet Sizes

Given that the bandwidth requirement of a VoIP stream can be minimized to about 10 kbps (e.g., through the use of a high-compression codec such as those discussed in Chapter 3), an 802.11b WLAN could, in principle, support hundreds of VoIP sessions. In reality, no more than a handful of sessions can be supported by an 802.11b WLAN due to various overheads.

As Figure 5.1 shows, the effective throughput in an 802.11b network has a large dependency on the payload size that is used. Even though this is not an issue for data applications (since they will most likely use large payload sizes), this does not bode well for VoWLAN where the packet size needs to be kept short to minimize end-to-end delay. As we discussed in Chapter 3, VoIP (and hence VoWLAN) uses packetization periods of the order of 10–40 ms leading to

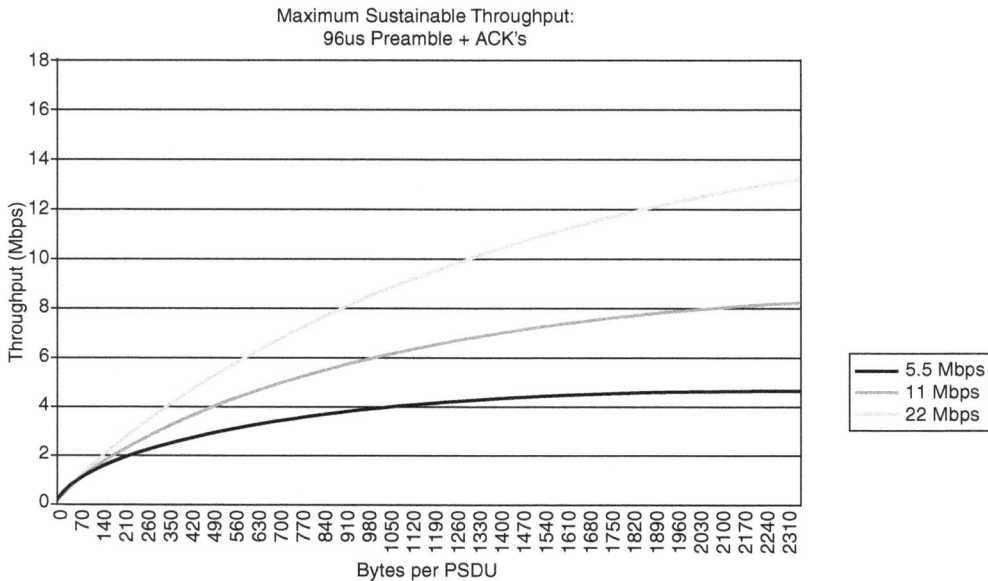

Figure 5.1: Throughput vs. Packet Size

payload sizes of the order of 100–300 bytes. As is clear from Figure 5.1, this limits the effective bandwidth available for VoWLAN to about 1–2 Mbps in a BSS.

Realize also from Figure 5.1 that using higher transmission rates helps improve system capacity, but this increase in system capacity is most significant at higher payload sizes and the gain at lower payloads is comparatively small. Applications like VoWLAN, which use small payloads, see only a small increase in system bandwidth since they lie in the bottom left corner of the graph.

To understand why the system capacity is a factor of the payload size, realize that the 802.11 MAC does not take into account the transmission time (for which the station would use the media once it captures it) when competing for media access. Instead, it concentrates only on making the number of transmission opportunities fair among stations. In other words, the MAC protocol ensures that, once a station gets access to the media and finishes its transmission, it must again compete with other stations to transmit its next packet. However, the MAC does not take into account how long the station would stay on the media once it gets access to it. So, once a station gets access to the channel, it may transmit a packet with a payload of 10 bytes or a payload of 2300 bytes; this difference is not taken into account when stations compete for access to the media. In effect, a station that transmits 2300-byte payloads on getting access to the media can pump much more data through the network than a station that transmits only 10 bytes of payload when it finally gains access to the media. Therefore, in VoWLAN systems, stations that use small payload sizes must spend a considerable amount of time backing off in the MAC protocol to avoid collisions, and this leads to limited system capacity.

5.3.2 Packetization Overheads

Another reason for the limited capacity of VoWLAN is the packet header overhead that is added as the short VoIP packets traverse the various layers of the standard protocol stack. The payload of a voice packet with a 10-ms packetization period, as generated by the voice codec, ranges from 10 to 80 bytes, depending on the codec used. This voice payload then passes down the stack via the RTP, UDP and IP layers. These three layers add headers of a total size of 40 bytes. Next, the IP layer hands over this packet to the 802.11 MAC protocol, which adds a header of 34 bytes. Note that, at this stage, the total packet size (assuming a 30-byte voice payload) is 104 bytes, out of which only 30 bytes is the actual voice payload. That is an efficiency of less than 30%.

Next, when the packet is handed over to the 802.11b PHY layer, a PLCP header and a PLCP preamble are added to it. Even though the size of these together is 15 bytes (short preamble) or 24 bytes (long preamble), the PHY overhead is significantly large since the transmission rate is limited to 1 or 2 Mbps. Assuming that the rest of the packet gets transmitted at the maximum 802.11b rate of 11 Mbps, this means that it takes 96 µs (short preamble) or 192 µs

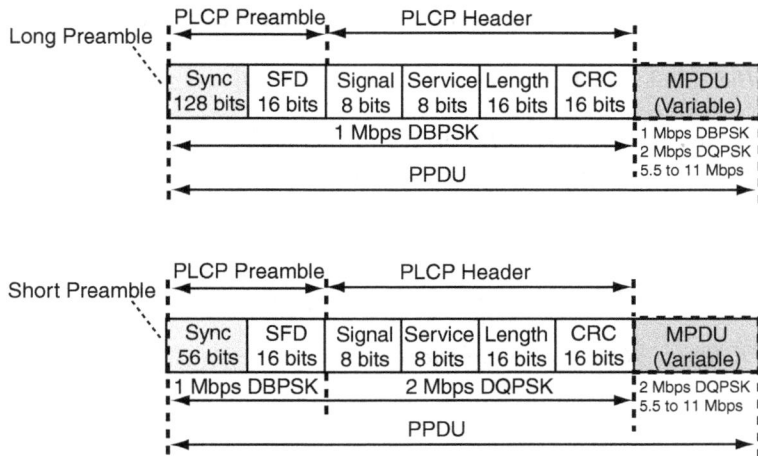

Figure 5.2: PHY Headers for 802.11b

(long preamble) just to transmit the PHY layer overheads. From a VoWLAN perspective, this means that to transmit 22 μs (30 bytes) of voice payload, it takes a total time of 172 μs (short preamble) or 268 μs (long preamble). That is an efficiency of about 9 to 13%, which means we are already down from 500 VoWLAN sessions to about 50 VoWLAN sessions in a BSS.

5.3.3 DCF Overheads

Recall from Chapter 4 that, in order to protect against nodes hogging the channel once they get access to it, the 802.11 MAC (DCF) requires that a station must wait between consecutive packet transmissions. This waiting period allows other stations to compete for channel access if needed and thus ensures that a station does not hog the channel once it gets access to it. However, this waiting period also means additional overheads. Let us calculate the time it takes to transmit a voice packet using DCF.

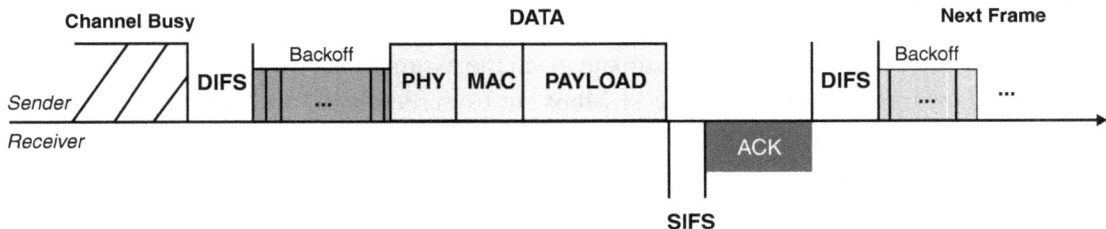

Figure 5.3: DCF Timing

From Figure 5.2, the total time it takes to transmit a voice packet can be calculated as:

Pkt_TxTime = DIFS + BO + PHY_TxTime + MAC_TxTime + Payload_TxTime + SIFS + ACK_TxTime.

For 802.11b, even in the best-case scenario (short preamble, maximum transmission rate, an aggressive BO time and a 30-byte payload packet), we have:

- *DIFS = 50 μs*

- *BO = Slot Time * CWavg = 20 * 31/2 = 310 μs assuming CWavg = (CWmin −1)/2*

- *PHY_TxTime = 96 μs assuming short preamble*

- *MAC_TxTime = 34 * 8/11=25 μs assuming the maximum*

- *Payload_TxTime = 70 * 8/11=51 uSec where "payload" includes RTP, UDP and IP headers*

- *SIFS = 10 μs*

- *ACK_TxTime = PHY_TxTime + (34 * 8/11) + (14*8/11) = 131 μs since the 14-byte ACK also comes with 802.11 MAC and PHY headers.*

- *Therefore Pkt_TxTime = 673 μs*

Continuing our calculations of the number of simultaneous voice calls in a BSS from section 5.3.2, from a VoWLAN perspective this means that to transmit 22 μs (30 bytes) of voice payload, it takes a total time of 673 μs, which reduces the efficiency to 3%.

Note the effect of ACKing each packet. The 802.11 MAC requires each data packet to be explicitly acknowledged (ACKed) to cope with operating in the (hostile) wireless environment. ACKing each packet also reduces the system capacity significantly.

5.3.4 Transmission Rate

Note that section 5.2.2 is a best-case estimate given the assumptions we made. For example, we assumed the transmission rate to be 11 Mbps but from section 4.3 we know that the transmission rate used is often a factor of channel conditions (which is also a factor of distance between the communicating stations). Transmitting at lower data rates would means that the transmission time for each packet increases and the system is occupied for more time, thus reducing system capacity even further for VoWLAN.

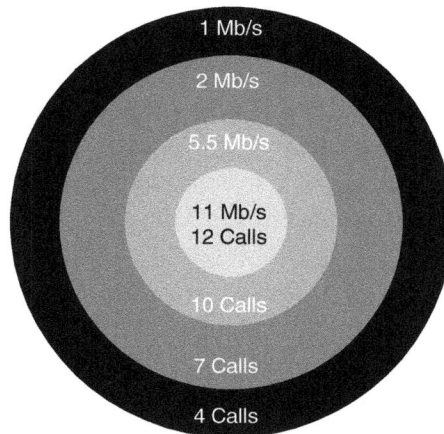

Figure 5.4: System Capacity vs. Range

Figure 5.4 uses a very simple radio channel model to illustrate the effect of transmission rate on system capacity in VoWLAN. We know that the strength of the signal decreases as the distance between the communication stations increases. In Figure 5.3, with the AP at the center of the figure, assuming a constant noise floor, the received signal strength (and hence the SNR) decreases as we move away from the AP. Therefore, the "optimum" transmission rate decreases as we move away from the AP. With a decrease in the transmission rate, the number of voice calls that can be supported also decreases as we move away from the AP. Note that Figure 5.3 is not drawn to scale and is for illustration purposes only.

Transmitting at higher data rates means that the transmission time for each packet decreases and the medium is freed up for use for more packets, thus increasing system capacity. However, transmitting at higher data rates means using more complex modulation schemes, which are more susceptible to channel noise. Therefore, using higher data rates in adverse channel conditions (high channel noise—i.e., lower SNR) can actually lead to higher BER—i.e., higher packet loss, which may require more retransmissions and thus effectively reduce system capacity.

The goal, therefore, is to dynamically adjust the transmission rate. Section 4.3 introduced this topic of rate adaptation in 802.11 networks. The concept of rate adaptation is to select the appropriate transmission rate based on channel conditions and performance. As also mentioned in section 4.3, the rate-adaptation algorithm is not specified in the 802.11 standards and this is expected to be one of the product differentiators among various vendors. It is important to realize that a rate-adaptation algorithm optimized for data may not yield the best results for real-time applications like voice. The primary reason for this is that voice is extremely sensitive to delay and jitter. We shall see how this affects the rate-adaptation algorithms.

In order to decide which rate is optimal at any specific moment, the rate-adaptation algorithm needs information about the current link conditions. Since it is difficult to get this information directly, most algorithms use some form of statistics-based feedback. The statistic most often used in such feedback schemes is the user-level throughput. This means that these algorithms aim to maximize the application-layer throughput. To achieve this, typical 802.11 rate-adaptation algorithms are "aggressive" in attempting to switch to a higher PHY data rate, the underlying theory being that if packet error rate increases at higher data rates, the 802.11 algorithm will cope with such drop-outs by using frame retransmissions. This approach works fine for data communication since the extra (and variable) delay introduced by this retransmission-dependent approach is acceptable to data applications. However, this increase in (average) packet delay and jitter (due to variations in the number of retransmissions) can cause serious degradation for voice communication. Consequently, rate-adaptation algorithms optimized for data communication often perform poorly for VoWLAN.

Realize also that there will be times in an 802.11 network where temporary network conditions will prevent the successful transmission of a packet under any rate adaptation. For example, the STA may have moved into an RF blind spot, or near a jamming device such as a microwave oven. In these cases, a voice-friendly rate-adaptation algorithm would want to give up on the current voice packet instead of delaying the entire voice packet stream trying to get the current voice packet through. Since VoIP packet-loss concealment algorithms can hide the loss of one or two packets but cannot mask a large drop-out, there is no point in taking pains to deliver a voice packet if it is so late that the receiver jitter buffer has underflowed. Again, the rate-adaptation algorithms are expected to be another product differentiator among vendors.

5.3.5 Inherent Fairness Among All Nodes

Besides the PHY and MAC overheads, there is another problem in VoWLAN that reduces the system capacity even further. An 802.11 BSS typically[2] consists of an AP and stations. The 802.11 standard specifies that both the AP and the station use the same MAC. Specifically, the AP and the stations use the same media contention/access scheme and back-off periods (contention window values). This means that if an AP and station both compete to get access to the channel, both are equally likely to get access to the medium. However, in a BSS, there is one AP and multiple stations. This means that if multiple stations and the AP compete for the media, it is more likely that one of the stations gets access to the media.

This inherent fairness among the stations and the AP works fine for data communication since most data applications[3] are highly asymmetrical—i.e., they have high download (from the

---

[2] In this book, VoWLAN refers to WLAN using the infrastructure BSS.
[3] A typical example is Internet browsing.

AP to the station) traffic and low uplink (from the station to the AP) traffic. Since stations rarely compete to access the media in data communication, the AP can easily get access to the media when it needs it.

However, voice communication is bidirectional and highly symmetrical; i.e., the downlink traffic and the uplink traffic are very similar in terms of bandwidth requirements. This means that stations need to access the media much more often than in data communication.

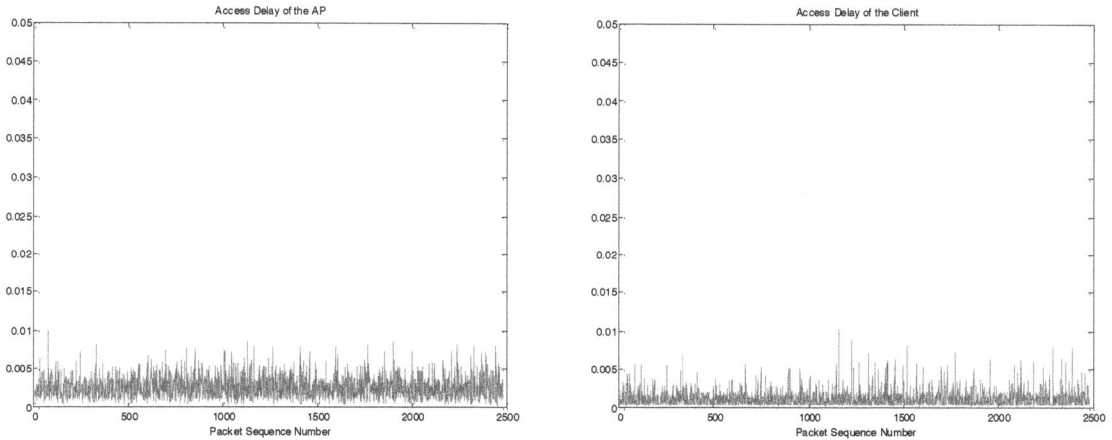

Access Delays in AP and a Station in Original VoIP over WLAN when there are 12 Sessions

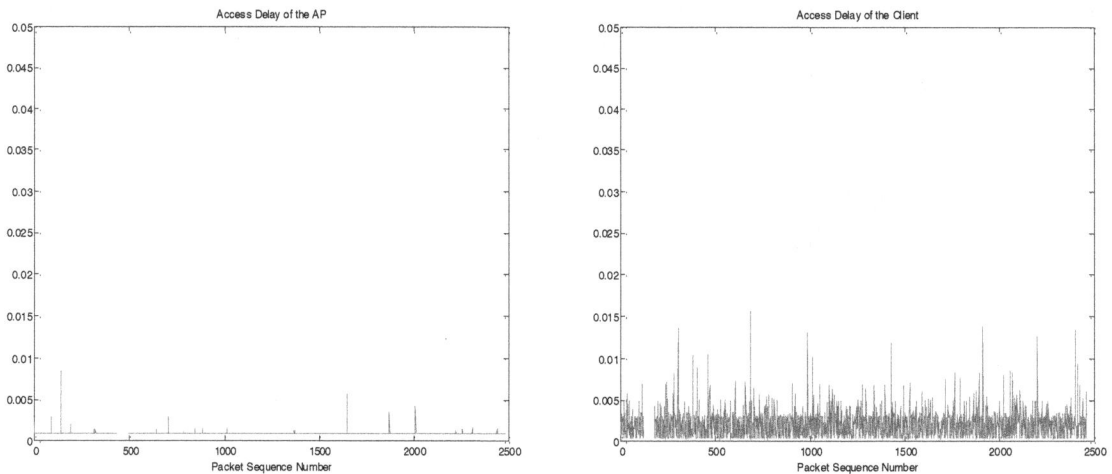

Access Delay in AP and a Station in Original VoIP over WLAN when there are 25 Sessions

Figure 5.5

Given that stations and AP are competing on an equal footing and that there are multiple stations involved in voice communication in a BSS, the AP is much less likely to get access to the channel than all other stations combined. Put another way, since the AP is as likely to get access to the channel as any other station, the probability of a station getting access to the media is higher than the probability of the AP getting access to the media if more than one station is competing for the media. Combine this with the fact that the AP is handling much more traffic[4] than any station and you have a system where the node handling the most traffic (AP) is not given priority over other nodes. This leads to a single point of back-up and congestion in a VoWLAN BSS as shown in Figure 5.5.[5]

In practice, if there are N wireless IP phones in a BSS making calls to wired networks, the AP is handling N times the load as compared to any other node in the BSS. However, fairness in 802.11 would allow the AP to access the medium only as much as any other node. The bottom line is that the AP will not be able to transmit the traffic that it is receiving. Bad, as this may sound, things get worse. Since we are dealing with real-time traffic, a packet which gets delayed beyond a limit waiting in the AP queue is rendered useless.

From the 802.11 MAC perspectives, this situation arises because 802.11 requires that every station that finishes a transmission AND has a packet waiting in its queue MUST perform the random back-off. In the build up to a congested network, the AP will almost always have more than one packet in its queue (since all packets must go through it) so it will be backing off. For an 802.11b network, this backing off will, on average, add a delay of 320 μs[6] to its effective packet-transmission time. All phones, however, in most cases will rarely have more than one packet to transmit unless congestion is really heavy or there are PHY layer problems (such as moving out of range of the AP).

5.3.6 Analysis

The previous sections have argued that the maximum number of voice calls that can exist in a 802.11b BSS is severely limited. Analysis[7] has quantized this observation as shown in Table 5.1.

---

[4] Since all traffic must pass through the AP, the AP is almost handling as much load as the combined load of all stations in the BSS.

[5] *A Multiplex-Multicast Scheme that Improves System Capacity of Voice-over-IP on Wireless LAN by 100%,* Wang, Liew et al.

[6] DIFS = 20 μs; $CW_{avg} = 32/2 = 16$; Delay = DIFS * CW_{avg}.

[7] *A Multiplex-Multicast Scheme that Improves System Capacity of Voice-over-IP on Wireless LAN by 100%,* Wang, Liew et al.

Table 5.1: VoIP Capacities for 802.11b

Codecs	Ordinary VoIP
GSM 6.10	11.2
G.711	10.2
G.723.1	17.2
G. 726-32	10.8
G. 729	11.4

This analysis has been confirmed by observation too. Put briefly, the number of VoIP connections that can exist in a BSS increases with the use of larger packetization periods. However, higher payload sizes mean larger end-to-end delays in VoIP. So, this increased system capacity comes at the cost of an increased end-to-end delay in VoWLAN systems. Hence, simply using very large packetization periods is not a viable solution for increasing system capacity in VoWLAN networks. We look at solutions to this problem in Chapter 6.

5.4 PCF

Section 5.3.1 explains that stations transmitting smaller packets are at a loss with respect to stations transmitting larger packets, since DCF (the default 802.11 MAC protocol) specifies the same back-off times for both regardless of the payload size they intend to use. Now, consider an 802.11 BSS where some stations are being used for VoWLAN communication whereas others are being used for data applications like Internet browsing. Since VoWLAN stations would be using smaller payload sizes, they are inherently at a loss when competing with data stations for access to the media. This situation is in fact the exact opposite of what is desired. Voice communication requires low delay and low jitter and it should therefore be given a higher priority. In 802.11 we have, in effect, voice traffic being given lower priority than data traffic. What is needed then is a way to prioritize voice over data traffic.

The IEEE 802.11 group were aware of the problems with using DCF for real-time traffic. They therefore defined the point coordination function (PCF) to be used with real-time traffic. The PCF divides time into superframes. Each superframe is further subdivided into a CFP (contention free period) followed by a CP (contention period). Therefore, in the PCF, CFP and CP alternate over time with the additional requirement that the CP be long enough to ensure the delivery of at least one MSDU. During the CP, channel access is still controlled as in DCF. However, in the CFP, channel access is controlled exclusively by the PC (point coordinator), which is a logical entity typically colocated at the AP.

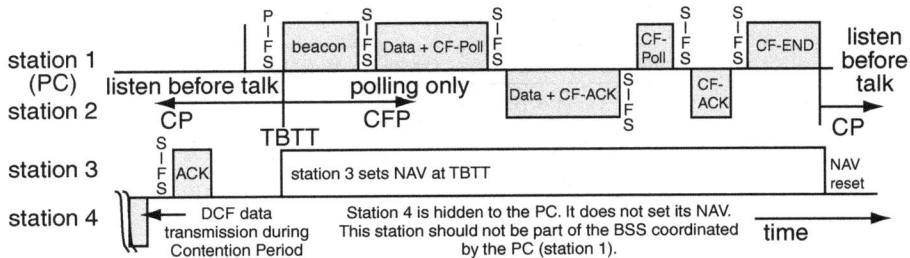

Figure 5.6: PCF Timing

The transmission of a beacon frame marks the beginning of a superframe. The beginning of the superframe also marks the beginning of the CFP. During this time, PC has complete and unilateral access to the channel and no other station may try to access it. In the CFP, the PC polls the stations for a chance to transmit data. A polled station is allowed to transmit a single MSDU when it is polled. This polling continues till the end of the CFP, which is indicated by the transmission of a CF-End packet. Ideally, the centralized tighter control of channel access by the PC should have allowed time-sensitive applications like voice to satisfy their QoS (delay and jitter requirements). However, there are many loopholes in PCF which make it unusable.

Consider what happens if the PC polls a station when the CFP is just about to end. Since the 802.11 standard allowed stations to start data transmission even if the MSDU delivery could not finish before the upcoming TBTT (target beacon transmission time), this basically meant that the CFP may actually end up the transmission of the next beacon. This unpredictability in TBTT stemming from the unknown transmission durations of the polled stations during a CFP is a serious limitation of PCF since it has serious repercussions not only for the QoS of other stations but also for power management in stations.

Consider another case where the PC polls a station and the station has an MSDU to transmit. Since the maximum allowed size of the MSDU is 2304 bytes, this station can effectively capture the channel for a long time, which would in turn delay the transmissions of other stations in the BSS. Note that this loophole stems from PCF specifying the polled station's channel access in terms of the number of MSDUs rather than absolute time. These problems have made PCF practically unusable and it has seen very little (if any) deployment in the industry.

5.5 Admission Control

Most networks require that a user/endpoint request permission to use the network before it actually starts using the network resources. This is known as admission control and can be broadly classified into two categories: authentication-based and resources-based.

Authentication-based admission control refers to a procedure in which the network ensures the authenticity of the user/endpoint before allowing it to access the network. This is usually done for the purposes of security and billing. The 802.11 standard allows for authentication-based admission control[8] where the network/AP may reject the admission request of a station if the station fails the authentication process.

Resource-based admission control has its roots in telephony networks like PSTN and is related to the concept of QoS. The underlying philosophy is that, if a network admits a user to the network, it must allocate/reserve resources for this user to ensure service to the user. In the case of telephony, we do not want the destination phone to ring if we can't ensure that resources for the call will be available (this is known as a call defect, and has a more stringent failure rate requirement in toll-quality phone networks). Since the network has limited resources, the network should keep track of what resources it is currently using and allow users to use the network only if there are enough resources available to service this user. Hence, admission control in this case is a matter of network capacity and usage. Since the 802.11 standard has its roots in data communications, where the network does not reserve any resources for any users, it does not provide any resource-based admission-control procedures. This, combined with the fact that the capacity of WLANs in terms of VoWLAN is extremely limited, means that congestion in VoWLAN networks is not handled gracefully. In most scenarios, a congested VoWLAN network severely degrades performance for all nodes in the system. A solution to this problem is also proposed in the 802.11e standard and is discussed in Chapter 6.

5.6 Security

Security in 802.11 networks is a complex issue and has been a focus of a lot of attention. Our concern here of course is that while we might not care if someone snoops to see what web pages we are downloading, we do care if someone can snoop on our conversations. We discuss 802.11 security and how it relates to VoWLAN in Chapter 7.

5.7 Power Save

Section 4.6 described the power save mechanism specified in the base 802.11 standard. To summarize, the 802.11 power save mechanism required that the AP buffer packets destined for a dozing station and inform the station about buffered packets in the beacon. This means that packets destined from the AP to the station may be delayed for as long as the beacon period. Since beacon periods are typically configured to be of the order of hundreds of milliseconds,[9] using this power-save approach for real-time traffic would lead to delays of

---

[8] The fact that this authentication scheme is weak and suffers from a lot of loopholes is another matter, which is discussed in detail in Chapter 7.

[9] Arguably, smaller beacon periods would make this approach suitable even for low-delay applications like VoWLAN; but smaller beacon periods also means that when a Wi-Fi handset is not in use, it will have to wake up more often, thus leading to more power consumption and lower battery lives. See Chapter 9 for more power-management issues.

hundreds of milliseconds in the voice path. Clearly, this is unacceptable. Hence, we conclude that the 802.11 doze mode is not suitable for real-time applications like VoWLAN.

5.8 Roaming/Handoffs in 802.11

The wireless medium is a harsh medium for signal propagation. Signals undergo a variety of alterations as they traverse the wireless medium. Some of these changes are due to the distance between the transmitter and the receiver, others are due to the physical environment of the propagation path, and yet others are due to the relative movement between the transmitter and the receiver.

Attenuation refers to the drop in signal strength as the signal propagates in any medium. All electromagnetic waves suffer from attenuation. For radio waves, if r is the distance of the receiver from the transmitter, the signal attenuation is typically modeled as $1/r^2$. It is important to emphasize that this is radio modeling we are talking about. Such models are used for simulation and analysis. In real life, radio propagation is much harsher and the signal strength and quality at any given point depend on a lot of other factors too. Attenuation of signal strength predicts the average signal strength at a given distance from the transmitter.

$$P_{RX} = P_{TX} \frac{G_{TX} \cdot G_{RX} \cdot \lambda^2}{16 \cdot \pi^2 \cdot d^2 \cdot L}$$ where:

G_{Tx} = transmitter antenna gain
G_{Rx} = receiver antenna gain
λ = wavelength (same units as d)
d = distance separating Tx and Rx antennas
L = system loss factor (≥ 1)

Attenuation explains why all wireless transmissions have a limited geographical range. The signal strength at the receiver decays as the distance between the transmitter and receiver increases. This decay in signal strength means receivers far away from the transmitter are more prone to suffer from transmission errors. To operate reliably in the harsh wireless medium, the 802.11 MAC protocol requires each packet to be individually ACKed by the receiver. If a packet is not ACKed, the transmitter must retransmit the packet. However, this protocol will work only up to a certain threshold distance, say R (range). If the distance between the transmitter and the receiver is increased beyond R, the received signal strength will be too low to achieve any communication. The value of R, therefore, depends upon:

- *Transmitted Power Level*: Increasing transmission power will increase the range. However, 802.11 operates in the unlicensed frequency spectrum and most governments restrict the maximum transmission power level in this spectrum to less than 1 watt.

- *Wavelength/Frequency*: Higher frequencies are more prone to attenuation than lower frequencies. Therefore 802.11b/g networks, which operate at 2.4 GHz, have a greater range than 802.11a networks, which operate at 5 GHz.

- *Antenna Gains*: Since transmitted power level and operating frequency are restricted by laws and standards, respectively, many 802.11 equipment manufacturers concentrate on antenna gains to increase their range. Most 802.11 equipment uses omnidirectional antennas since they are designed for general-purpose networking. However, it is possible to custom-design WLAN products per specific requirements to increase the range of 802.11 networks. Use of intelligent antennae and MIMO (multiple in multiple output) is also getting considerable attention in the industry for this purpose.

The bottom line, however, is that an 802.11 BSA has limited geographical range, typically a few hundred feet. Given that mobility is an inherent expectation in wireless networking, the question is how to provide seamless connectivity to a mobile user in 802.11 networks. This is where roaming comes in. When a station moves out of the range of its AP and enters into the range of another AP, a handoff[10] is said to have occurred.

It is important to realize that an 802.11 handoff is a Layer-2 process. An analogy might help make things clearer. Consider a laptop which is plugged into a corporate LAN socket using an Ethernet cable. This laptop has a valid IP address, which applications use to access the intranet and the Internet. Now, if the Ethernet cable is plugged out from its current LAN socket and plugged into another socket belonging to the same corporate LAN, the applications on the laptop can continue to use its IP address. In fact, the applications are not even aware of what has happened. The 802.11 handoff is a similar process where the plugging in and out of the Ethernet cable from sockets is analogous to disconnecting from an AP and connecting to another AP.

Therefore, if a higher-layer application uses a reliable transport-layer protocol like TCP, it does not even need to be aware of the 802.11 handoff—the transport layer will take care of retransmitting packets that were lost during the handoff. A web browser running HTTP (HyperText Transfer Protocol) is such an example. Consider a mobile user running a web browser on his laptop which connects to the Internet via an 802.11 network. When the user clicks on an HTTP link while a handoff is in progress, the underlying transport protocol, TCP, conceals the delay/packet loss due to the handoff by using retransmissions. In other words, higher layers (HTTP, the web browser and the user) are unaware of the handoff and can continue without disruption. On the other hand, if a higher-layer application uses an unreliable transport layer protocol like UDP, it would "see" the handoff as temporarily increased delay or packet loss. As long as the application can tolerate and recover from this delay or loss, an 802.11 handoff would be "transparent" to the user too.

---

[10] The terms handoff and roaming are used interchangeably in 802.11 networks.

The challenge for VoWLAN is that voice is extremely sensitive to delay. The end-to-end delay budget for voice is 250 ms; this means that the accumulative delay between the two endpoints involved in a voice call must not exceed 250 ms. This 250 ms must include the total transmission delay, propagation delay, processing delays in the network and and codec delays at both endpoints. In WLANs, the 802.11 MAC introduces an extra transmission delay in the WLAN due to the contentious nature of the MAC protocol.[11] This wireless transmission delay increases as networks become more congested or suffer from interference. The bottom line is that the budget for each component of the accumulative delay will typically be specified by the service provider, but we can't assume that we would have all 250 ms available for 802.11 delays. A good VoWLAN implementation would aim to keep the 802.11 delays limited to 40–50 ms.

Given that VoWLAN operates on a very restricted delay budget, the 802.11 handoff times on the order of a few hundred milliseconds[12] are unacceptable for voice. Thus, handoff times are another important area of product differentiation and vendors are competing to minimize handoff times in their products. We look at roaming in detail in Chapter 8.

5.9 Summary

In this chapter we looked at the major challenges facing VoWLAN, including QoS and capacity, power, security and roaming. In subsequent chapters, we will discuss proposed solutions that aim to tackle these challenges and make VoWLAN ready for wide-scale deployment.

---

[11] The situation is worse if both endpoints in the voice call use VoWLAN.
[12] Typically between 200 and 500 ms.

QoS and System Capacity

6.1 Introduction

Section 5.3 described the challenges of system capacity and QoS in some detail. This chapter focuses on the solutions to these challenges. As discussed in section 5.3, the DCF (distributed coordination function) defined in the base 802.11 standard was designed primarily for handling data traffic. When used for voice traffic, it ends up giving lower priority to voice than to data, which is disastrous for voice communication. Furthermore, the PCF (point coordination function) defined by the 802.11 standard for real-time communication such as voice suffered from some major drawbacks (section 5.3.1) which made it practically unusable. In order to make WLAN suitable for voice communication, the IEEE standards body created the 802.11e working group to come up with a solution.

Meanwhile the Wi-Fi Alliance, which is basically an industry consortium created for ensuring interoperability between various WLAN vendors, did not want to wait for the 802.11e standard since it was taking a long time to ratify. In the interest of getting products to the market faster, the Wi-Fi Alliance took a snapshot of the 802.11e standard as is and froze it to formalize the WME (WLAN Multimedia Enhancement) standard. As the WME specification says:

> *"WME ... (is) motivated by the need to prevent market fragmentation caused by multiple non-interoperable pre-standard subsets of the draft 802.11e standard that would otherwise occur. .... In no way should WME be taken to detract from 802.11e itself, which is viewed as the long term endpoint of WME."*

The WME standard is therefore a parallel (though very similar) standard to 802.11e to achieve QoS in WLANs. Some primary differences between the two standards are that WME does not include support for HCF,[1] block acknowledgments and associated signaling. Some of these terms will make more sense at the end of this chapter. We will point out the significant differences between the WME and 802.11e standards as we go along.

---

[1] Hybrid Coordination Function (HCF), however, is now included in WMM-SA (Wi-Fi MultiMedia-Scheduled Access) which is the follow-up standard of WME from the Wi-Fi Alliance.

6.2 802.11e, WME and "Vanilla" WLANs

802.11 networks have seen widespread deployment in the past few years. Most of these networks were deployed primarily for data networking. As VoWLAN gains popularity, solutions like 802.11e and WME are expected to see increased deployment. However, given the widespread deployment of WLANs, an overnight change from vanilla (deployments with no QoS support) WLANs to 802.11e WLANs cannot be expected. Instead most networks are expected to undergo a transition path where some devices in the network support 802.11e, some support WME, others support WMM-SA and yet others would be vanilla (support no QoS). Therefore, an important challenge is to be able to determine which device supports what. Towards this end we first distinguish between APs and stations.

An AP can indicate its support for QoS by carrying the appropriate IE (information element) in its beacons and Probe Responses. This allows stations in the BSS to determine at association time whether or not they can use QoS mechanisms. Note that the IEs used for 802.11e and WME are different, thus allowing an AP to support both 802.11e and WME simultaneously. Figure 6.1 shows the WME information element carried by beacons from a WME-capable AP. The presence of this IE in the beacon is then an indication to the stations in the BSS that this AP is capable of supporting WME.

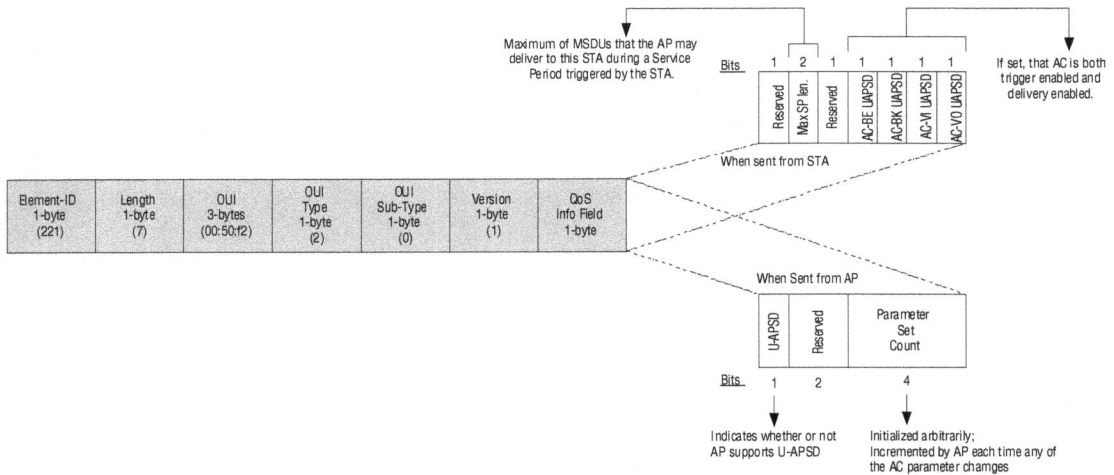

Figure 6.1: WME Information Element

Now, given that an AP can indicate its support for QoS by adding the appropriate IEs in its beacon frames, the next question is how does the station indicate which QoS standard (if any) the station wishes to use. The station indicates this preference again by using the appropriate IE in its Association Request.

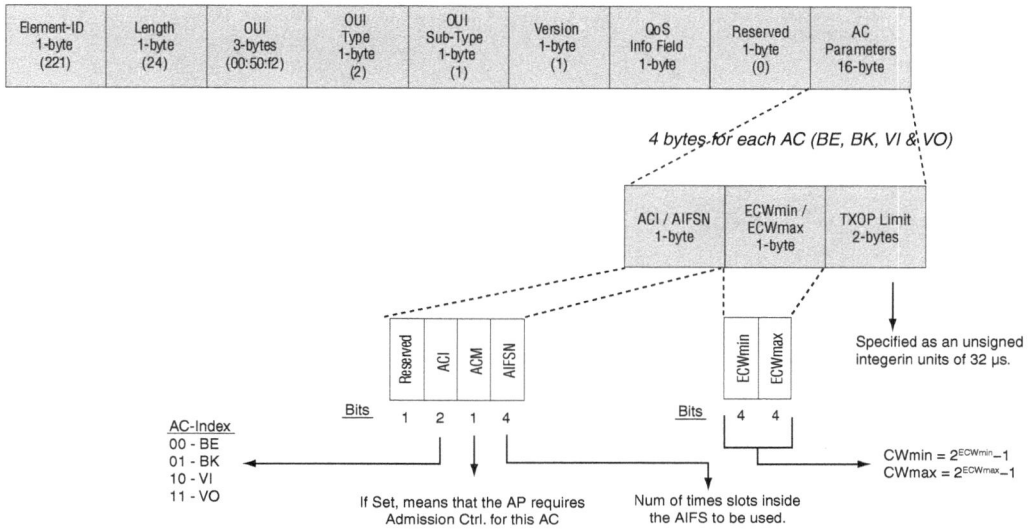

Figure 6.2: WMM Parameter Element

If the Association Request from the station contains a WME information element (like the one shown in Figure 6.1), the AP responds with an Association Response containing a WME parameter element (see Figure 6.2). This association is then a WME association. On the other hand, if the Association Request from a station does not contain a QoS IE, this means that the station does not support (or does not want to use) any QoS for this association. Therefore, the Association Response in this case does not contain any QoS IE.

Finally, since different stations in a BSS may have different QoS capabilities, what is also needed is a way to distinguish between QoS and non-QoS 802.11 frames. This is achieved using the type-subtype fields in the Frame Control field of the MAC header, as shown in Figure 6.3

Note that QoS frames (i.e., frames transmitted from a QoS station) use a 26-byte MAC header instead of the standard 24-byte MAC header. The two extra bytes are used to carry the QoS control field as shown in Figure 6.3. Now, that we have a basic understanding of the QoS frame types, in the next few sections we will look into how QoS is achieved using 802.11e/WME.

May be used to set NAV so as to protect bursts
of MSDUs when using continuation TXOPs

Present if and only if frame is of subtype
QoS data or QoS Null

| Frame Control 2-bytes | Duration Id 2-bytes | Add-1 (Receiver) 6-bytes | Add-2 (Transmitter) 6-bytes | Add-3 (BSSID) 6-bytes | Sequence Control 2-bytes | Add-4 (BSSID) 6-bytes | QoS Control 2-bytes | Body / Data / Payload n-bytes | Frame Check Seq. 2-bytes |

protocol	type	sub-type	to-DS	from-DS	More-frag	Retry	Pwr Mng	More-Data	WEP	Order
2	2	4	1	1	1	1	1	1	1	1

Seq. Num.	Frag. Num.
12	4

Reserved	ACK-Policy	EOSP	Reserved	User-Pri
9	2	1	1	3

Sequence number maintained
on a per-AC basis

New for QoS
Type 10; SubType 1000 - QoS Data
Type 10; SubType 1100 - QoS Null
Type 00; SubType 1101 - Management Action

Expected ACK response:
00 - Acknowledge
01 - Do Not Acknowledge

802.1D priority Tag

End-Of-Service Period: for U-AP SD;
Set to 1 by AP to indicate the end of a Service Period

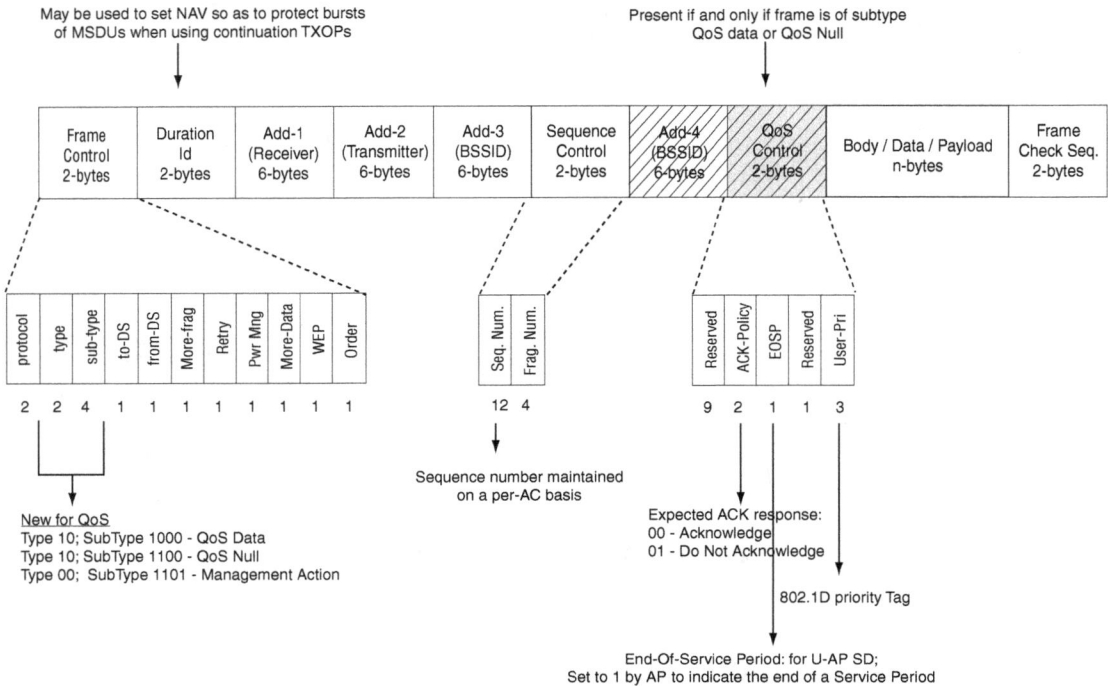

Figure 6.3: 802.11 QOS Frame Format

6.3 Traffic Categories

As a first step to providing QoS, 802.11e classifies traffic into categories according to the requirements of the traffic. As we know, voice traffic has very different requirements from data traffic. Voice traffic wants minimal delay but can tolerate certain levels of loss due to the inherent redundancy in voice. On the other hand, data traffic does not tolerate loss at all, whereas it is much less concerned with delay. Classification of traffic is therefore a good first step towards implementing QoS.

This classification into traffic categories (TCs) is then exploited by using multiple queues in a WLAN device as shown in Figure 6.4. In the legacy 802.11 standard, all traffic (regardless of whether it was voice, data or video) from a station was transmitted from a single queue. This meant that (small) voice packets could be delayed during bursts of data transmission. Realize that, since data packets are much larger than voice packets and since data traffic is often bursty in nature, this could mean a delay of several hundreds of milliseconds for voice packets, thus rendering them unusable. With the use of multiple queues, as proposed in 802.11e, separate queues are maintained for voice and data.

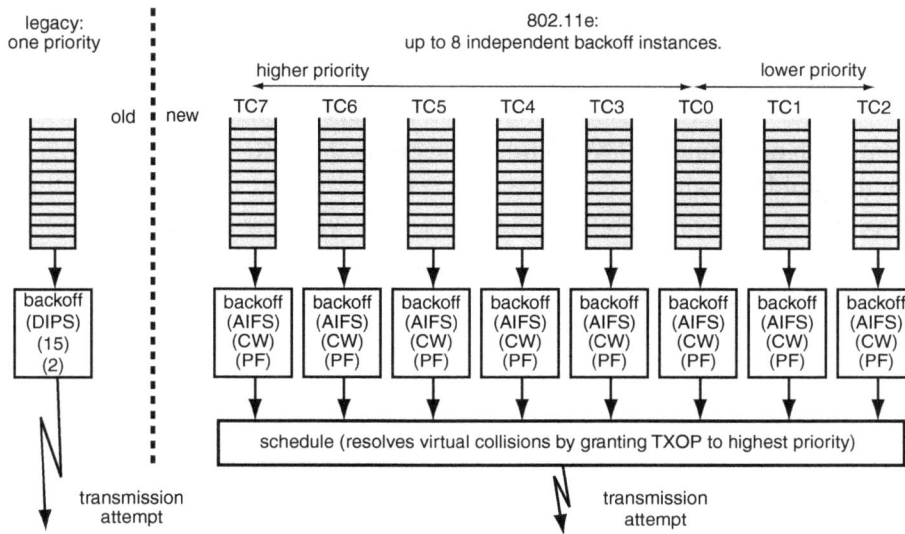

Figure 6.4: 802.11e Traffic Categories (TC)

With this basic queuing framework in place, a simple priority mechanism could then be used to ensure that packets in the higher-priority (voice) queue are transmitted before packets in the low-priority (data) queue. However, a strict priority scheme like this one could potentially starve off data packets. Therefore, 802.11e proposes that contention among various queues inside a station be resolved in the same way as the contention between multiple stations—i.e., by using one of the new coordination functions proposed in WME, using EDCF. (We look at the Enhanced DCF (EDCF) in section 6.5.) However, to summarize, the basic idea of traffic categories is that each station maintain multiple (up to eight) queues for different traffic categories. Each queue then acts as a virtual station and these virtual stations compete among themselves to select which packet would be transmitted from the station. The rules of competition among these queues are the same as the MAC rules used among stations for accessing the channel.

Note that 802.11e allows a station to classify its traffic in up to eight different traffic categories, where each TC can be mapped to a separate queue. However, WME restricts the number of queues in a station to four and explicitly specifies the mapping between 802.1D priority tags and the TCs.

6.4 Transmission Opportunity

802.11e introduces the concept of a transmission opportunity (TXOP). A TXOP is defined as an interval of time when a station has the right to initiate transmission, defined by a starting time and a maximum duration. A TXOP is allocated either via contention (as in EDCF) or granted through polling (as in HCF). The maximum duration of a TXOP is a BSS-wide limit, which is announced/distributed by the AP in beacon frames when using EDCF and is announced in the poll frame when using HCF.

On one hand, a station cannot transmit a frame that extends beyond a time interval called EDCF transmission opportunity (TXOP) limit;[2] this is used by HCF (see section 6.6) to resolve one of the problems with PCF. On the other hand, a station can transmit multiple MAC frames consecutively (with an SIFS time gap between an ACK and the subsequent frame) as long as the whole transmission time does not exceed the EDCF TXOP. This mechanism is known as contention-free burst (CFB).

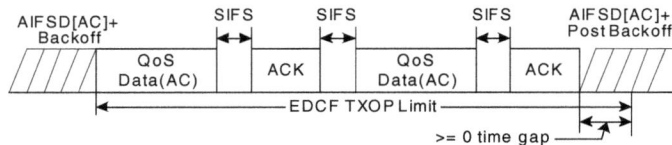

Figure 6.5: 802.11 QOS Data Frame Transmission During EDCF

Figure 6.5[3] shows the transmission of two QoS data frames during an EDCF (see section 6.5) TXOP, where the whole transmission time for two data and ACK frames is less than the EDCF TXOP limit announced by the AP.

The use of CFBs is not suitable per se for voice communication, since voice packets are periodic in nature. However, in mixed environments where voice, video and data coexist, CFB increases the system throughput without degrading other system-performance measures unacceptably, as long as the EDCF TXOP limit value is properly determined. This is shown in Figure 6.6.

---

[2] If a frame is too long to be transmitted in a single TXOP, it should be fragmented into multiple frames.
[3] Taken from IEEE 802.11e *Contention Based Channel Access (EDCF) Performance Evaluation,* Choi, Prado et al.

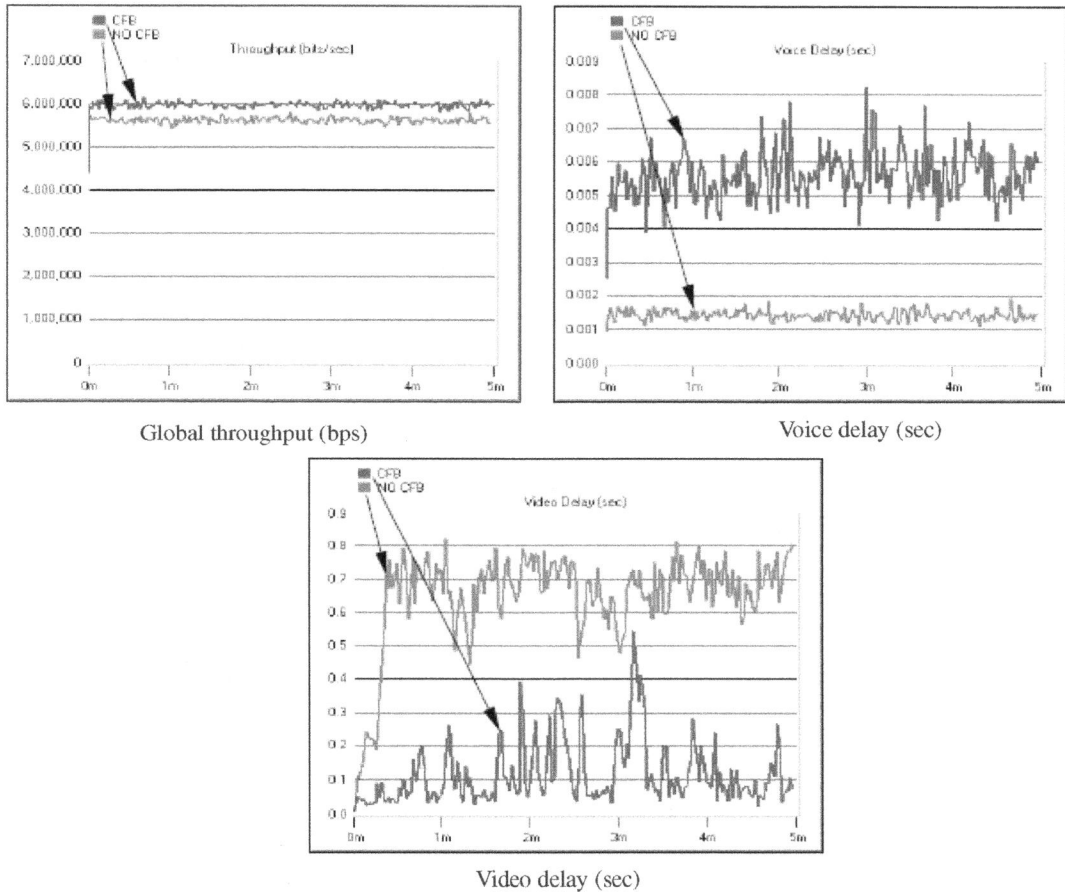

Global throughput (bps)

Voice delay (sec)

Video delay (sec)

Comparison of EDCF with and without CFB.

Figure 6.6: EDCF Performance with and without CFB

6.5 EDCF

The EDCF (Enhanced Distributed Coordination Function) protocol is, as the name suggests, an enhanced version of the legacy DCF protocol of 802.11. The EDCF protocol has been adopted as WME (Wireless Multimedia Enhancement) by the Wi-Fi Alliance as the prestandard implementation of 802.11e.

The aim of the EDCF protocol is to ensure prioritized access to the (wireless) channel so that traffic in the voice TC can access the channel with a higher priority than traffic in the background or best-effort (data) TC. Note that whether the queues are inside one station or inside two different stations is irrelevant to the protocol. The aim is to ensure that packets in the higher priority (voice) queues get prioritized access to the channel.

EDCF achieves this aim by maintaining contention parameters on a per-TC basis. The term "contention parameters" refers to the following set of parameters used by the 802.11 MAC:

Vanilla 802.11	WME / 802.11e
CWmin	*CWmin [TC]*
CWmax	*CWmax[TC]*
DIFS	*AIFS [TC]*
PF	*PF [TC]*

To understand how maintaining these parameters on a per-TC basis achieves prioritized access for different types of traffic, it is instinctive to recall some salient features of DCF from section 4.5.5. In any moderately loaded BSS,[4] when a station wants to get access to transmit on the wireless channel, it waits till the channel is idle. After sensing the channel idle for the first time, the station waits for a minimum period of DIFS. After the DIFS expires, each station waits for an additional random time determined by a back-off counter (BC). The BC value is randomly selected from the range [0, CW] where CW refers to the contention window.

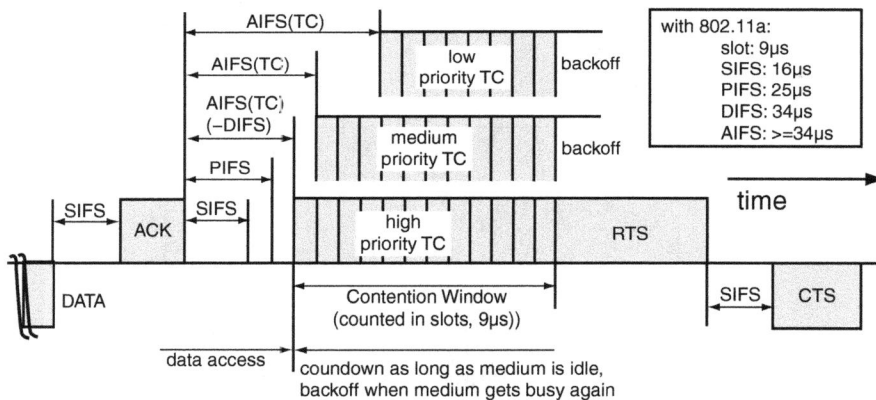

Figure 6.7: EDCF Operation

First, in EDCF, the minimum time to wait after sensing the channel is made TC-dependent. Instead of all traffic having to wait for DIFS before accessing the medium, higher-priority (voice) TCs have to wait a smaller time AIFS[TC$_{voice}$] than lower-priority (data) TCs, as shown in Figure 6.7. Second, EDCF allows each queue to maintain a different set of the contention window limits. Again, higher-priority (voice) TCs have a smaller range of CW limits than lower-priority (data) TCs. Together, this means that if a voice and a data packet compete

---

4 In very lightly loaded systems, when the station senses that the medium has been longer than DIFS time, it can begin transmitting frames immediately.

for access to the channel, the voice packet is more likely to get access to the channel, thus creating a prioritized access to the medium.

Finally, recall from section 4.5.5 that an unsuccessful (un-ACKed) transmission of an MSDU led to the doubling of CW, thus leading to potentially longer back-off times for retransmission of a packet. These longer back-off times affected time-sensitive (voice) traffic more adversely than data. To solve this problem, 802.11e introduced the concept of persistence factor (PF). The PF, which is specified on a per-TC basis, determines how long each queue backs up before attempting a retransmission. For legacy DCF, PF was always 2, since CW was doubled after unsuccessful transmission. In 802.11e, the PF is used to calculate the new CW in case of transmission failures as follows: *newCW = ((oldCW + 1) * PF) −1*. By using a lower PF for higher-priority (voice) queues, voice packets can be allowed to attempt retransmission faster, thus ensuring lower delays even for retransmitted voice packets.

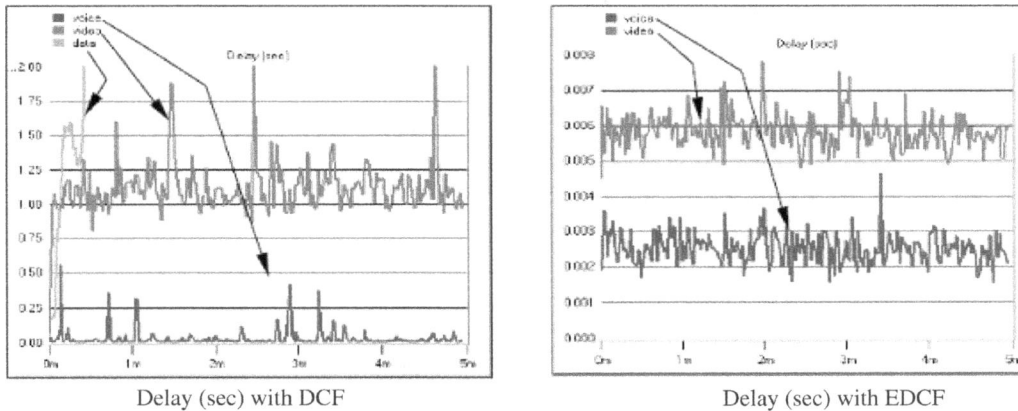

Delay (sec) with DCF   Delay (sec) with EDCF

THREE TRAFFIC TYPES AND CHARACTERISTICS

Type	Inter-arrival Time (Avg. in sec)	Frame Size (bytes)	Data Rate (Mbps)
Voice	Constant (0.02)	92	0.0368
Video	Constant (0.001)	1464	1.4
Data	Exponential (0.012)	1500	1.0

EDCF PARAMETERS USED FOR SIMULATIONS

Type	Prior.	AC	AIFSD	CWmin	CWmax	TXOP limit (msec)
Voice	7	3	PIFS	7	15	3
Video	5	2	PIFS	15	31	5
Data	0	0	DIFS	31	1023	0

Figure 6.8: EDCF vs. DCF Performance

Thus, the CW limits, AIFS and PF work together to achieve prioritized access for voice in 802.11e/WME. Figure 6.8[5] shows the advantage of using EDCF over DCF in voice communication. The data used for these figures was obtained in a scenario simulating four voice stations, two video stations and four data stations. Note that each delay curve is from a single station. Also note that the scale on the *y*-axis for the two graphs is different.

Besides improving the delay performance for voice, EDCF also improves system performance in mixed environments where voice, video and data co-exist. Figure 6.9[6] shows system throughput and data-dropped performance comparision between DCF and EDCF. Specifically, video traffic gains significantly with the use of EDCF.

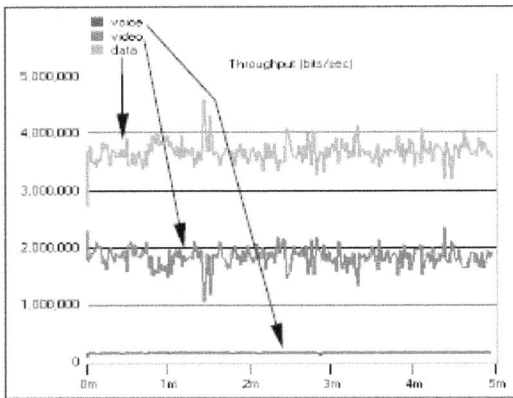

Throughput (bps) with DCF

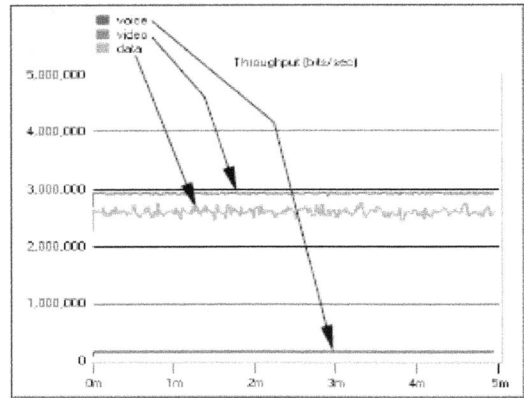

Throughput (bps) with EDCF

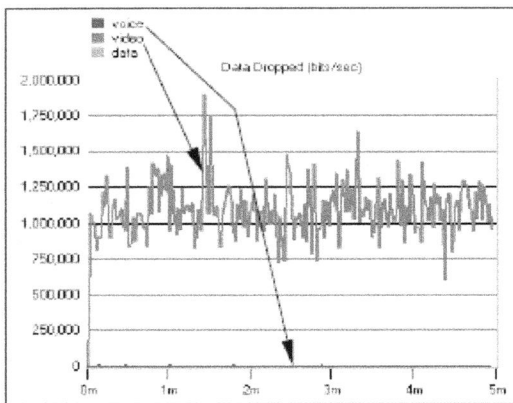

Data dropped (bps) with DCF

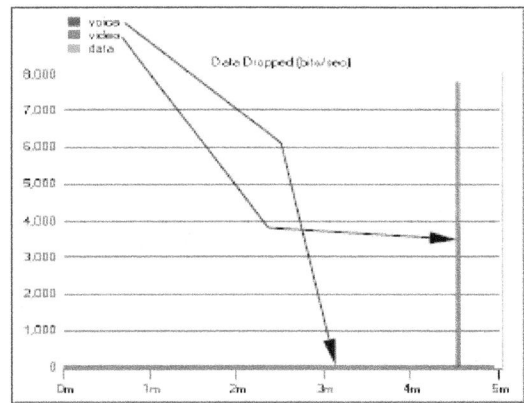

Data dropped (bps) with EDCF

Figure 6.9: EDCF vs. DCF Performance in Voice/Data/Video Environments

---

[5] Taken from IEEE 802.11e *Contention Based Channel Access (EDCF) Performance Evaluation*, Choi, Prado et al.
[6] Ibid.

Note that the value of QoS parameters for each TC is announced by the AP in the WME parameter element (see Figure 6.2), which is carried in the Association Response, Probe Response and/or beacon. This ensures that all stations in the BSS—except the AP—are treated equally. This is an important exception. Recall from section 5.3.5 that one of the problems that led to the limited system capacity and lack of QoS in the original 802.11 standard was the inherent fairness among all nodes in the system where the AP had to contend, on an equal footing, with stations in the BSS. 802.11e/WME allows the AP to use a different set of contention parameters than those used by the stations. This means that the AP can get access to the medium at a higher priority than stations in the BSS, which solves the problem of the AP being a bottleneck in a VoWLAN system.

6.6 HCF

Just like EDCF was adopted in the WME standard by the Wi-Fi Alliance, HCF was adopted in the WMM-SA (Wi-Fi MultiMedia-Scheduled Access) specification by the Wi-Fi Alliance as a successor to WME. It is instinctive to think of the HCF (hybrid coordination function) as a successor to PCF, just as EDCF is a successor to DCF. EDCF is meant to enhance the DCF feature and make it QoS-compliant, where HCF is meant to enhance the PCF feature and make it suitable for QoS.

The primary difference between EDCF and HCF is that the former provides prioritized QoS and the latter provides parameterized QoS. Prioritized access refers to real-time traffic (voice and video) getting a higher priority when contending for access to the channel. Parameterized access, on the other hand, refers to promising predefined delay limits for real-time traffic. As we shall see, HCF achieves this by using a polling-based approach just like PCF. However, it is important to realize that a polling-based approach brings with it the added overhead of polling frames, which consumes bandwidth. Therefore, an EDCF mechanism (which does not consume bandwidth in polling frames) would perform well under conditions of light to moderate loads whereas an HCF-based mechanism would perform well under heavy loads. Network administrators of VoWLAN systems must keep this in mind when deciding upon a choice between EDCF and HCF.

HCF, as we said, can be thought of as the enhanced version of PCF. Recall from section 5.4 that one of the primary problems of PCF was that, once a station got access to the channel, its access was not constrained in time (it could hog the channel). The TXOP (see section 6.4) solves this problem by limiting the transmission opportunity in time; so even when a station gains access to the channel using the MAC rules, the station has limited time for which it can access the channel. The use of TXOP avoids the unpredictable delays that are possible in PCF. The transmission opportunity is therefore the maximum contiguous time a station can use the channel once it gets access to it. Note that during a TXOP, a station is allowed to transmit multiple MSDUs with an SIFS gap between an ACK and the subsequent frame as long as the station honors the TXOP limit.

HCF uses the EDCF protocol as a building block and extends the concepts of CFPs, CPs and polling from PCF. Just like PCF, time is divided into superframes. Each superframe consists of a CP and a CFP. Only the HC (located at the AP) has access to the channel during the CFP and it polls stations for granting access to them during this time. During CP, the EDCF rules are used to decide access to the channel.

Unlike PCF, however, HCF allows the HC to poll stations even during a CP. Therefore, during a CP, a TXOP begins either when the medium is determined to be available under the EDCF rules or when the station receives the special CF-POLL frame.

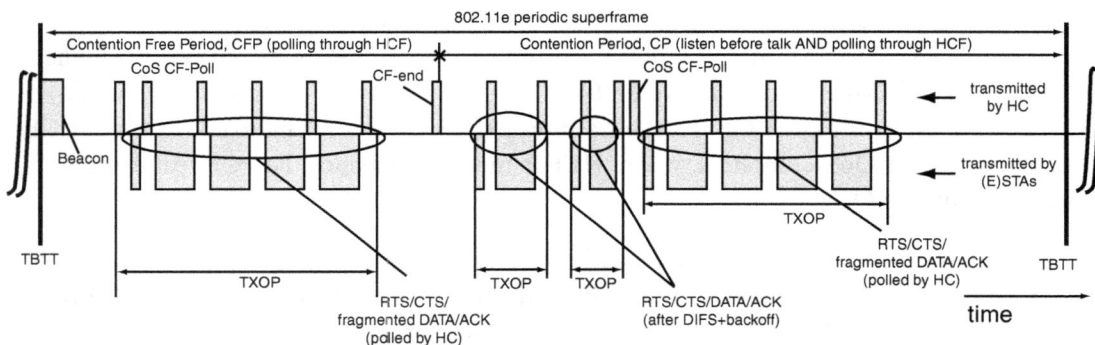

Figure 6.10: WMM-SA Operations

Note that the use of TXOP resolves the problem of unpredictable delays. In HCF, the TXOP may be specified explicitly in the CF-poll, which is sent to the station to grant it access to the channel. A station is expected to give up access to the channel within this TXOP.

One issue with HCF is that it requires a well-managed radio spectrum in order to be effective. With HCF, the packet scheduler in the AP will be the crucial component. This is not defined in the specification (although approaches are suggested). However a real-life scheduler will need to to account for interference from other channels or other networks sharing the same spectrum, unless the spectrum is completely controlled. This will complicate the scheduler considerably. Thus, we should expect that HCF will be more of an enterprise solution and less likely to be used in home networks.

6.7 Voice Data Coexistence

The aim of 802.11e was to enhance the standard to make it suitable for voice communication. However, it is important to remember that the 802.11 standard was originally designed for data communication. Hence, it is imperative that the 802.11e standard should not negatively affect data communication; 802.11e WLANs must support both data and voice communication.

EDCF achieves this by using probabilities to achieve QoS in media access instead of a strict priority scheme, as strict priority scheme among the multiple queues would have given higher priority to voice packets at the risk of starving off data packets. In a WLAN consisting of both voice and data stations, this could lead to voice stations hogging the wireless channel and data stations starving off. By using lower values for CW limits, AIFS and PF for voice queues, EDCF ensures that when voice packets and data packets compete for access to the wireless medium, voice packets have a higher probability of getting access to the channel than data packets. Note that EDCF deals with probabilities. This means that even though voice packets are more likely to get access to the channel, they cannot hog the wireless channel since data packets will also sometimes win the contention to access the channel.

Similarly, HCF ensures that data packets are not starved off by using transmission opportunities and by ensuring a contention period in each superframe. Together these ensure that a voice station or voice packets do not hog the frame and starve off data transmission.

6.8 Achieving QoS for VoWLAN

It is important to realize that 802.11e/WME provides a mechanism for 802.11 devices to prioritize voice packets over data packets in the wireless LAN. However, these mechanisms alone are not sufficient for achieving QoS for VoWLAN calls since a VoWLAN call may (and usually will) extend beyond the WLAN. This is an extremely important point to realize, since it is often a source of confusion. Simply using 802.11e/WME is not sufficient for achieving QoS. To understand why this is so, consider a wireless IP phone (WIPP) in a voice call with a wired IP phone (IPP) located on the far side of an IP Network. For purposes of explanation, Figure 6.11 divides end-to-end QoS into three segments: the wireless LAN, the wired LAN and the IP Network. We now consider what is needed to achieve QoS in each of these segments.

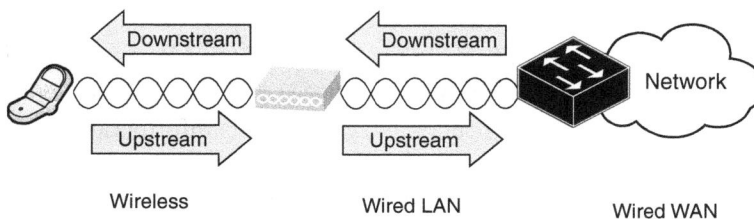

Figure 6.11: Wi-Fi Phone to Network Voice Flow

6.8.1 Wireless LAN

In the wireless upstream, 802.11e/WME achieves QoS since stations competing to transmit packets to the AP would give higher priority to voice packets than data packets.

At first, achieving QoS in the downstream may seem trivial since the AP is also capable of using 802.11e/WME to prioritize its traffic over the wireless medium. However, there is an important piece of information missing in the wireless downstream—the AP needs to be able to distinguish between voice packets (both signaling and media) and data packets to actually prioritize one over the other. In the upstream, the application running on the station is responsible for classifying packets into the correct TC. In our example, the voice application on the WIPP may somehow inform the WLAN driver of which packets belong to which category. However, in the downstream, the AP is handling packets from a remote endpoint. Therefore, it needs to be able to classify downstream packets by examining the packets. This is where it gets interesting. Realize that the AP is a bridge (i.e., a Layer-2 device) intended to ensure the interworking between 802.3 and 802.11. Therefore, as per the OSI layered model, it should access only the Layer-2 headers. In other words, what we need is a field in the 802.3 (Layer-2) header for the packets arriving on the wired downstream at the AP that can be examined by the AP to classify the packets into the correct TC. This is discussed in section 6.8.2.

6.8.2 Wired LAN

Probably the most widely deployed mechanism to achieve QoS in the wired LAN is the use of 802.1D/Q. The IEEE 802.1D MAC bridge specification allows different MAC layers in the IEEE 802 family to interwork. The 802.1Q Virtual LAN (VLAN) tag extends the existing 802.3 frame format, and it specifies the user priority of the frame. Note that the 802.3 MAC itself does not support any differentiated channel access to different priority traffic, but via the 802.1Q VLAN tag, the 802.3 MAC frames can carry the corresponding priority value, which in turn can be used by the 802.1D MAC bridge for prioritized forwarding.

If the AP supports both 802.11e and 802.1D/Q, it can be used to achieve QoS in a LAN environment. To continue with our example, when such an AP receives an Ethernet packet with a 802.1D/Q header, it can examine the priority bits in the 802.1Q header to decide whether to transmit the packet from the AP's voice queue or the BE queue in the wireless network.

From a system perspective, however, there is still a missing part of the puzzle: who inserts the 802.1Q priority bits? Refer to Figure 6.11. In the upstream, the AP can distinguish between the voice packets and data packets by examining the 802.1D priority field in the QoS Control field of the 802.11e / WME MAC header and can use this information to set the 802.1Q priority bits in the 802.1D header that it attaches around the 802.3 header. Similarly, in the downstream, the packets received at the wired AP port must already have the 802.1Q priority bits in the 802.1D part of the 802.3 header so that the AP can map them to the 802.11e header. How the packets arriving at the AP's wired port get these priority bits forms the topic of the next section.

6.8.3 IP Network

When packets transmitted from the IPP destined for the WIPP reach the AP, there is no absolute way for the AP to know that these are voice packets. Note that the IPP setting the 802.1Q/D priority bits in the MAC header is not an option, since Layer-2 (MAC) headers are changed on a hop-by-hop basis and the IPP and WIPP can potentially be separated by multiple hops. Since the AP cannot distinguish between voice and data packets in the wired downstream by looking at the 802.3 header, it cannot decide which packets need to be transmitted from the AP's voice queue and which need to be transmitted from the data (BE) queue. The bottom line is that to achieve QoS in the wireless downstream, 802.11e needs other mechanisms. This is where Layer-3 QoS becomes involved, since Layer-3 headers are maintained end-to-end.

DiffServ is probably the most widely deployed Layer-3 QoS protocol. In this approach, each IP datagram carries a *differentiated service code point* (DSCP) value in its differentiated service (DS) field, which supercedes the IP V4 type of service (TOS) octet and IP V6 traffic classifier octet. Since IP is a Layer-3 (end-to-end) protocol, the IP header is available end-to-end and can be used by routers in the IP Network to provide differentiated service i.e., QoS.

Now, consider our example of a wireless IP phone (WIPP) in a voice call with a wired IP phone (IPP). Irrespective of whether the two are separated by an IP Network or whether they belong to the same LAN, if they decide to use DiffServ, the DSCP value can be used to distinguish voice packets from data packets in the network as is done by routers supporting DiffServ in the IP Network.

When the packet reaches wired LAN, 802.1D/Q may be used if needed to provide QoS in the wired LAN. However, this may not be required if the Ethernet is a 100-Mbps or a Gigabit Ethernet. The important thing to note is that this approach does not rely on 802.1D/Q headers to provide QoS in the wireless domain.

When a DiffServ packet reaches the AP, the AP may look inside the IP header to read the DSCP and then map[7] the packet to the appropriate 802.11e queue. In effect, the AP uses the DSCP to distinguish between voice and data packets. Strictly speaking, this approach is a violation of the layered architecture since the DSCP value is carried in the Layer-3 IP header, whereas an AP is a Layer-2 device. In a strict layered architecture, a Layer-2 device (the AP) should not be "looking at" the Layer-3 header (the DSCP). However, to make 802.11e inter-work with DiffServ, that is exactly what is required.

---

[7] The mapping between DSCP and 802.11e can be made configurable and left up to the network administrator for better control.

Using this approach and assuming that QoS in the wired LAN is not required, in our example of a wireless IP phone (WIPP) in a voice call with a wired IP phone (IPP), both of them will set the DSCP field in the IP header. For voice packets traveling from the WIPP to the IPP (via the AP), nothing special is required since 802.11e provides for QoS from the WIPP to the AP and the DSCP value inserted by WIPP can be used to provide QoS in the WAN. For voice packets traveling from the IPP to the WIPP (via the AP), the DSCP value is used to provide QoS in the WAN and the AP maps the DSCP value to the 802.11e priority, thus providing QoS in the wireless downstream.

6.8.4 LAN-only QoS

Note that the approach discussed in 6.8.3 does not need to be restricted to cases where a WAN is involved. Setting DiffServ within a LAN environment is also possible—and probably more consistent from a deployment perspective.

6.9 System Capacity

We have seen how 802.11e/WME aims to give higher priority to voice traffic over data traffic by using the TCs and enhanced MAC algorithms. These features help in reducing the delay encountered by voice traffic, thus enhancing the quality of communication. An additional advantage of these enhancements is that they increase the system capacity for Voice-over-WLANs. To see how, recall from section 5.2 that the capacity of a WLAN to carry voice calls is limited to a small number due to the following factors:

a. Small size of voice packets.

b. High Physical layer overheads.

c. Voice and data compete "equally" for access to media.

d. Stations and AP compete "equally" for access to media.

e. Requirement of a per-packet ACK.

First, let us start with what 802.11e/WME does not do. It does not solve problem (a) which is essentially a feature/limitation of VoIP. Voice packets are kept small to minimize delay and this is something that VoWLAN must live with.

Table 6.1: Overhead Parameters in Various 802.11 PHY Layers

	802.11b	802.11g + b	802.11g-only	802.11a
DIFS (μs)	50	50	28	34
SIFS (μs)	10	10	10	16
Slot Time (μs)	20	20	9	9
CWmin	32	16	16	16
Supported Data Rates (Mbps)	1,2,5.5,11	1,2,5.5,11, 6,9,12,18,24,36,48,54	6,9,12,18,24,36,48,54	6,9,12,18,24,36,48,54
Basic Rate (Mbps)	2	2	N.A.	N.A.
PHY for protection frames (μs)	N.A.	192	N.A.	N.A.
PHY for other frames (μs)	192	20	20	20
ACK frame (μs)	131	24	24	24

802.11e/WME does not solve problem (b) either, which is a PHY layer problem. The solution to (b) lies in choosing the right PHY layer for VoWLAN. As Table 6.1 shows, using 802.11g-only[8] or 802.11a as the PHY-layer is much more suitable for VoWLAN since it significantly reduces the PHY layer overheads. This is verified by analysis[9] too, as shown in Table 6.2.

Table 6.2: VoIP Capacities for 802.11b, 802.11a and 802.11g Derived from Analysis

MAC	Ordinary VoIP
802.11b	11.2
802.11a	56.4
802.11g-only	60.5
802.11g with CTS-to-self protection	18.9
802.11g with RTS-CTS protection	12.7

The 802.11e/WME enhancements to the problem of system capacity concentrate first and foremost on the inherent fairness in DCF (i.e., problem (c) above). Recall from section 5.2.2 that DCF concentrates on making the number of transmission opportunities fair among stations. In other words, it ensures that once a station gets access to the media and finishes its transmission, it must again compete with other stations to transmit its next packet. However, this is unfair to smaller voice packets since it does not take into account how long the station

---

[8] Since 802.11g operates in the 2.4-GHz band, it is possible to have a WLAN where the AP supports both 802.11g and 802.11b and some stations in the BSS are 802.11g whereas others are 802.11b. This leads to an 802.11g+b deployment, which reduces the effective system capacity of the system.

[9] *A Multiplex-Multicast Scheme that Improves System Capacity of Voice-over-IP on Wireless LAN by 100%*, Wang, Liew et al.

would stay on the media once it gets access to it. 802.11e fixes this by allowing (smaller) voice packets with a higher probability (and therefore more often) than (larger) data packets. This means that voice stations that use smaller payload sizes spend less time backing off than data packet, which is actually "fair" since they occupy the channel for less time when they do in fact get access to it.

Another aspect of fairness in the base 802.11 standard stems from the fact that it specifies that both the AP and the station use the same MAC parameters (problem (d) above). This means that if an AP and station both compete to get access to the channel, both are equally likely to get access to the medium. However, in a Voice-over-WLAN system, due to the bidirectional and asymmetrical nature of voice, the AP is handling much more traffic[10] than any station and we have a system where the node handling the most traffic (AP) is not given priority over other nodes. This leads to a single point of congestion in a VoWLAN BSS. EDCF fixes this problem by allowing the AP to use a different set of contention parameters (CW limits, AIFS and PF) from the station. This allows the AP a higher prioritized access to the wireless medium, which is suitable for voice communication. HCF, on the other hand, fixes this problem by giving complete control of medium access to the AP during the contention free period (during which voice communication is scheduled). The AP can therefore use the channel more often than the station for voice streams.

Finally, with respect to (e), 802.11e makes ACKs optional, thus allowing (at least theoretically) the number of packets over the air-interface to be reduce by half. However, given the harsh wireless environment, not using ACKs is not a realistic solution. However, what can be exploited is the Block-ACK extension of this feature where ACKs of N packets may be sent together. Sending one ACK for N packets instead of an ACK for every packet is expected to reduce the number of packets over the air interface and thus help improve system capacity. However, experiments have shown that, given the harsh wireless environment in which WLANs often operate, it is often best to ACK every packet.

Even though 802.11e/WME helps improve system capacity in terms of the number of simultaneously active calls in a BSS, the bottom line is that there are only a certain number of voice calls (factor of the voice packet size, etc.) that a BSS can support at any given time. Once the number of calls crosses this threshold, the BSS is said to be congested. A congested BSS leads to degradation in the voice quality of all voice calls that are active. Obviously, this is not a desirable scenario. What is needed is some form of an admission control where no more than a certain number of calls are allowed to be simultaneously active. 802.11e provides for TSPECs to implement admission control.

---

[10] Since all traffic must pass through the AP, the AP is almost handling as much load as the combined load of all stations in the BSS.

6.10 Admission Control

The need for admission control with Voice-over-Wi-Fi stems from the fact that there are always a finite number of voice calls that a BSS can support simultaneously. Once this threshold number of calls is exceeded, it would lead to degradation in the voice quality of all voice calls that are active in the BSS. Since this is not a desirable situation, what is needed is a mechanism to prevent the number of calls in a BSS to exceed this threshold.

Towards this end, 802.11e/WME provides for admission-control mechanisms using TSPECs (traffic specifications). The underlying philosophy of this approach is that a station does not have the right to use the medium for data transmission/reception simply because it is connected to or associated with the AP. Instead, a station must explicitly ask for permission (from the AP) before it can send or receive data. The exact mechanism is as follows. When a station needs to start data transmission/reception, it issues a TSPEC request to the AP. A TSPEC request is a new management frame introduced by 802.11e/WME (see Figure 6.8), which specifies exactly the properties of the packet stream that the station expects to transmit/receive by specifying the following important information about the packet stream:

a. Nominal MSDU size – For voice this would be based on the codecs negotiated for the call.

b. Mean Data Rate – For voice this would be based on the nominal MSDU size and packetization period used for the call.

c. Minimum PHY rate.

d. Surplus bandwidth allowance.

e. Medium time.

For a voice call, a TSPEC request should happen as early in the call setup as possible so as to avoid the situation where the called party can pick up a ringing phone to discover that the call cannot go through. It is preferable that the calling party be prevented from contacting the destination at all if there is insufficient bandwidth.

However, with VoIP, the media characteristics may not be finalized until the end of the call signaling procedure. For example, with SIP, the called party may not indicate his codec preferences until the 200 OK message is sent. A possible workaround for this case is for the calling party to issue a TSPEC with the worst-case call parameters that it would expect to use for the call. After the call parameters are negotiated and a lower-bit-rate codec is selected for the call, a [nice] VoIP station can reissue the TSPEC requesting less capacity.

0 - ADDTS request
1 - ADDTS response
2 - DELTS
3–255 - Reserved

0 - Admission Accepted
1 - Invalid Parameters
2 - Reserved
3 - Admission Refused
4–255 - Reserved

Response Status Codes for ADDTS

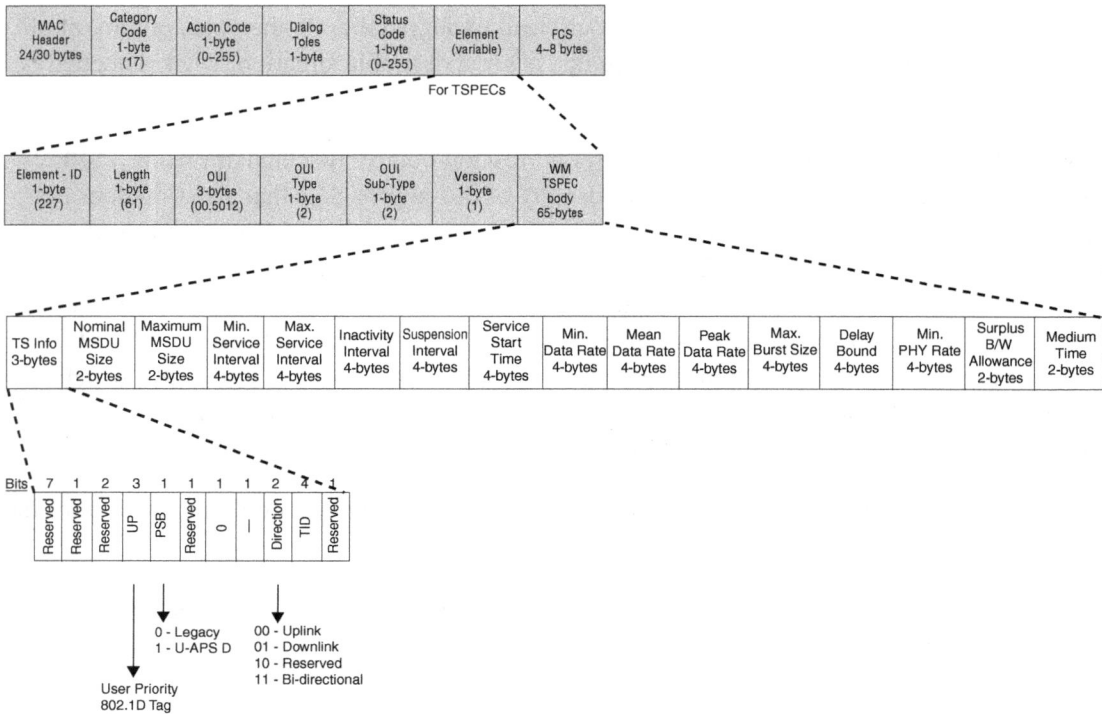

MAC Header 24/30 bytes	Category Code 1-byte (17)	Action Code 1-byte (0–255)	Dialog Toles 1-byte	Status Code 1-byte (0–255)	Element (variable)	FCS 4–8 bytes

For TSPECs

Element - ID 1-byte (227)	Length 1-byte (61)	OUI 3-bytes (00.5012)	OUI Type 1-byte (2)	OUI Sub-Type 1-byte (2)	Version 1-byte (1)	WM TSPEC body 65-bytes

TS Info 3-bytes	Nominal MSDU Size 2-bytes	Maximum MSDU Size 2-bytes	Min. Service Interval 4-bytes	Max. Service Interval 4-bytes	Inactivity Interval 4-bytes	Suspension Interval 4-bytes	Service Start Time 4-bytes	Min. Data Rate 4-bytes	Mean Data Rate 4-bytes	Peak Data Rate 4-bytes	Max. Burst Size 4-bytes	Delay Bound 4-bytes	Min. PHY Rate 4-bytes	Surplus B/W Allowance 2-bytes	Medium Time 2-bytes

Bits | 7 | 1 | 2 | 3 | 1 | 1 | 1 | 1 | 2 | 4 - 1

Reserved | Reserved | Reserved | UP | PSB | Reserved | 0 | I | Direction | TID | Reserved

0 - Legacy
1 - U-APS D

00 - Uplink
01 - Downlink
10 - Reserved
11 - Bi-directional

User Priority
802.1D Tag

Figure 6.12: WMM TSPEC Message Format

The AP uses the information provided in the TSPEC request to determine if the addition of traffic from this proposed session would lead to congestion or otherwise negatively affect the existing sessions or calls. If the AP decides that it can handle this traffic session, it sends a TSPEC response back to the station accepting its request and allowing the station to proceed with the call. If, on the other hand, the AP determines that this new session or call should not be allowed into the BSS, it will send a TSPEC response rejecting the station's TSPEC request.

Note that the algorithm used by the AP to decide whether or not to accept a TSPEC request is not a part of the standard and each vendor is free to use its own version. This decision is based on a host of factors including theoretical maximum number of voice calls possible, existing number of voice calls, bandwidth needed to be reserved for handling roaming calls,[11]

---

[11] An AP may decide to reserve some bandwidth to handle voice traffic from stations that may roam into its range while actively involved in a voice call.

data sessions in the BSS, interference in the BSS, etc. This is another area of product differentiation among vendors.

6.10.1 Traffic Categories and Admission Control

Recall from section 6.1.1 that 802.11e/WME classifies traffic into categories according to the requirements[12] of the traffic. 802.11e allows a station to classify its traffic in up to eight different traffic categories (TCs) and each station maintains a separate queue for each traffic category. The TSPEC mechanism used for admission control also operates on a per-TC basis—i.e., admissions are requested and granted on a per-TC basis. Furthermore, an AP may require admission control for some TCs and not for others. In fact, it is expected that most Voice-over-WLAN deployments would require admission control in the voice TC (to ensure QoS) but not in the best-effort TC (to allow for nonvoice traffic from the station).

The information as to which TCs the AP requires admission control for is contained in the beacons and Probe Responses transmitted by the AP. When a station has data to transmit, it must first determine which TC it wishes to use to transmit this data. Then, depending on whether or not the AP requires admission control for this TC, the station may need to send a TSPEC request and wait for a TSPEC response from the AP. If the AP accepts the TSPEC request, the station can then start the new stream. The interesting situation, however, is when the AP rejects (for whatever reason) the TSPEC request.

6.10.2 Handling Rejected TSPECs

Since the algorithm used by the AP to decide whether or not to accept a TSPEC request is not a part of the standard, an AP may reject a station's admission request (TSPEC) for various reasons. In fact, the TSPEC response sent by the AP to the station contains a reason field that can be used to carry the reason a TSPEC request is being rejected. Since the AP may reject the TSPEC request for various proprietary reasons, the value of this reason field may also be proprietary. It is expected that vendors providing a complete VoWLAN solution (i.e., AP, stations, etc.) will exploit this feature for product differentiation. For example, one way to do this is to integrate admission control, roaming and the voice-signaling stack. To understand the advantage of combining these different components, consider some of the reasons an AP may reject a TSPEC request from a station.

The probable primary reason for this would be that the BSS is congested and the AP determines that it would not be able to handle any more voice streams. In such a situation, the station has various options to handle the rejected TSPEC from the AP. The trivial solution

---

[12] Voice traffic wants minimal delay but can tolerate certain levels of loss due to the inherent redundancy in voice. On the other hand, data traffic does not tolerate loss at all, whereas it is much less concerned with delay.

is to "propagate" the rejected TSPEC response to the voice-signaling stack so that the voice call that is being attempted can be gracefully stopped. This handling of a rejected TSPEC is analogous to a user picking up a PSTN phone and not getting a dial tone (or getting a fast busy tone).

Another option for handling a TSPEC rejected due to congestion in a BSS is for the station to attempt roaming to another available BSS in the same ESS. This is the basic idea of load balancing, where a station roams from one AP to another, not due to RSSI variations but because of system-capacity utilizations.

Yet another option for handling a TSPEC rejected due to congestion in a BSS is for the voice-signaling stack in the station to reattempt the admission control for the voice call with a different set of call parameters, primary among which are the voice codec and the size of voice packets, aka the packetization period. This approach is useful because using larger packet sizes for voice reduces the "loading" on the BSS by reducing the number of packets per second. This is reflected in the reattempted TSPEC request's MSDU size and mean data rate. The AP may then find these new parameters more acceptable than those in the original TSPEC request. Note that this would involve a VoIP call-signaling renegotiation step; for example, with SIP a new Invite message would need to be sent.

The bottom line is that vendors providing a complete VoWLAN solution (i.e., AP, stations, etc.) will exploit the flexibility in TSPECs for product differentiation.

6.10.3 Some Issues With TSPECs

One issue with the TSPEC message format is that it does not contain an "urgent" field. This would provide a mechanism for a Wi-Fi phone to indicate that it was making a priority—e.g., 911—call. This would let the AP prioritize its bandwidth requests (and even terminate bandwidth to an existing call) when 911 calls are active. To add such a field would be fairly easy; the real issue is how to authenticate the use of this field to prevent improper use.

Another issue with the standard is that a TSPEC authorization does not carry over the entire BSS. Thus, a Wi-Fi phone that needs to roam to a new AP will have to reissue the TSPEC. Furthermore, the TSPEC cannot be contained in an Association Request. This means that another message request/response latency will be imposed on the roaming timeline. We will discuss this problem further in Chapter 8.

6.11 Summary

This chapter looked at how 802.11e, WME and WMM-SA provide solutions for implementing QoS and improving system capacity for VoWLAN. In the next chapter, we tackle security for VoWLAN.

Security

7.1 Introduction

As we have stressed throughout this book, 802.11 is a Layer-2 standard. The OSI Layer 2 is concerned with establishing link-layer connectivity; that is, connecting devices to each other. Again, Layer-2 protocols traditionally have not had any security built into them because in the wired world physical access to the media is a big deterrent at Layer 2. To understand this, consider a local area network (LAN) designed using 802.3 (Ethernet). This LAN would consist of the physical wires, switches and hubs. To connect to such a LAN, a user would typically have to physically "plug-in" his computer to a switch or a hub. In other words, a user needs physical access to the network to connect to it. This requirement of having physical access to the network to connect to it serves as a security mechanism at Layer 2. However, this inherent security mechanism is missing in wireless networks.

The base 802.11 security architecture/protocol is called WEP (*Wired Equivalent Privacy*). It is responsible for providing authentication, confidentiality and data integrity in 802.11 networks. To understand the nomenclature, realize that 802.11 was designed as a "wireless Ethernet." The aim of the WEP designers was therefore to provide the same degree of security as is available in traditional wired (Ethernet) networks. Did they succeed in achieving this goal?

A few years back, asking that question in the wireless community was a sure way of starting a huge debate. To understand the debate, realize that wired Ethernet[1] (i.e., the IEEE 802.3 standard) implements no security mechanism in hardware or software. However, wired Ethernet networks are inherently "secured" since the access to the medium (wires) that carries the data can be restricted or secured. On the other hand in "wireless Ethernet" (i.e., the IEEE 802.11 standard) there is no provision to restrict access to the (wireless) media. So, the debate was over the fact whether the security provided by WEP (the security mechanism specified by 802.11) was comparable to ("as secure as") the security provided by restricting access to the physical medium in the wired Ethernet. Since this comparison is subjective, it was difficult to

---

[1] We use 802.3 as a standard of comparison since it is the most widely deployed LAN standard. The analogy holds true for most other LAN standards, more or less.

answer this question. In the absence of quantitative data for comparison, the debate raged on. However, recent loopholes discovered in WEP have pretty much settled the debate, concluding that WEP fails to achieve its goals.

In this chapter, we look at WEP, why it fails and what is being done to close these loopholes. It is interesting to compare the security architecture in 802.11 with the security architecture in TWNs (Traditional Wireless Networks—i.e., voice-based wireless networks like GSM). Note that both TWNs and 802.11 use the wireless medium only in the access network—i.e., the part of the network which connects the enduser to the network. This part of the network is also referred to as the *last hop* of the network. However, there are important architectural differences between TWNs and 802.11.

The aim of TWNs was to allow a wireless subscriber to communicate with any other wireless or wired subscriber anywhere in the world while supporting seamless roaming over large geographical areas. The scope of the TWNs, therefore, went beyond the wireless access network and well into the wired network.

On the other hand, the aim of 802.11 is only last-hop wireless connectivity. 802.11 does not deal with end-to-end connectivity. In fact, IP-based data networks (for which 802.11 was initially designed) do not have any concept of end-to-end connectivity and each packet is independently routed. Also, the geographical coverage of the wireless-access network in 802.11 is significantly less than the geographical coverage of the wireless-access network in TWNs. Finally, 802.11 has only limited support for roaming. For all these reasons, the scope of 802.11 is restricted to the wireless-access network only. As we go along in this chapter, it would be helpful to keep these similarities and differences in mind.

7.2 Key Establishment in 802.11

The key establishment protocol of 802.11 is very simple to describe—there is none. 802.11 relies on "pre-shared" keys between the mobile nodes/stations (henceforth STAs) and the APs. It does not specify how the keys are established and assumes that this is achieved in some "out-of-band" fashion. In other words, key establishment is outside the scope of WEP.

7.2.1 What's Wrong?

Key establishment is one of the toughest problems in network security. By not specifying a key-establishment protocol, it seems that the WEP designers were sidestepping the issue. To be fair to WEP designers, they did a pretty good job with the standard. The widespread acceptance of this technology is a testament to this. In retrospect, security was one of the issues where the standard did have many loopholes but, then again, everyone has perfect vision in hindsight. Back to our issue, the absence of any key-management protocol led to multiple problems, as we discuss below.

a.   In the absence of any key-management protocol, real-life deployment of 802.11 networks ended up using manual configuration of keys into all STAs and the AP that wished to form a BSS.

b.   Manual intervention meant that this approach was open to manual error.

c.   Most people cannot be expected to choose a "strong" key. In fact, most humans would probably choose a key that is easy to remember. A quick survey of the 802.11 networks that I have access to shows that people use keys like "abcd1234" or "12345778" or "22222222," etc. These keys, being alphanumeric in nature, are easy to guess and do not exploit the whole key space.

d.   There is no way for each STA to be assigned a unique key. Instead, all STAs and the AP are configured with the same key. As we will see in section 7.4.4, this means that the AP has no way of uniquely identifying a STA in a secure fashion. Instead the STAs are divided into two groups. Group One consists of stations that are allowed access to the network and Group Two consists of all other stations (i.e., STAs that are not allowed to access the network). Stations in Group One share a secret key that stations in Group Two don't know.

e.   To be fair, 802.11 does allow each STA (and AP) in a BSS to be configured with four different keys. Each STA can use any one of the four keys when establishing a connection with the AP. This feature may therefore be used to divide STAs in a BSS into four groups, if each group uses one of these keys. This allows the AP a little finer control over reliable STA recognition.

f.   In practice, most real-life deployments of 802.11 use the same key across BSSs over the whole ESS.[2] This makes roaming easier and faster, since an ESS has many more STAs than a BSS. In terms of key usage, this means that the same key is shared by even more STAs. Besides being a security loophole to authentication (see section 7.4.4), this higher exposure makes the key more susceptible to compromise.

7.3 Anonymity in 802.11

We saw that subscriber–subscriber anonymity was a major concern in TWNs. Recall that TWNs evolved from the voice world (i.e., the PSTN). In data networks (a large percentage of which use IP as the underlying technology), subscriber anonymity is not such a major concern. To understand why this is so, we need to understand some of the underlying architectural differences between TWNs and IP-based data networks. As we saw in Chapter 1,

---

[2]   Recall that an ESS is a set of APs connected by a distribution system (like Ethernet).

TWNs use IMSI for call routing. The corresponding role in IP-based networks is fulfilled by the IP address. However, unlike the IMSI, the IP address is not permanently mapped to a subscriber. In other words, given the IMSI, it is trivial to determine the identity of the subscriber. However, given the IP address, it is extremely difficult to determine the identity of the subscriber. This difficulty arises because of two reasons. First, IP addresses are often dynamically assigned using protocols like DHCP (*Dynamic Host Configuration Protocol*). That is, the IP address assigned to a subscriber can change over time.

Second, the widespread use of NAT (*Network Address Translation*) adds another layer of identity protection. NAT was introduced to deal with the shortage of IP addresses.[3] It provides IP-level access between hosts at a site (LAN) and the rest of the Internet without requiring each host at the site to have a globally unique IP address. NAT achieves this by requiring the site to have a single connection to the global Internet and at least one globally valid IP address (hereafter referred to as GIP). The address GIP is assigned to the *NAT translator* (aka NAT box), which is basically a router that connects the site to the Internet. All datagrams coming into and going out of the site must pass through the NAT box. The NAT box replaces the source address in each outgoing datagram with GIP and replaces the destination address in each incoming datagram with the private address of the correct host. From the view of any host external to the site (LAN), all datagrams come from the same GIP (the one assigned to the NAT box). There is no way for an external host to determine which of the many hosts at a site a datagram came from. Thus, the usage of NAT adds another layer of identity-protection in IP networks.

7.4 Authentication in 802.11

Before we start discussing the details of authentication in 802.11 networks, recall that the concepts of authentication and access control are very closely linked. To be precise, one of the primary uses of authentication is to control access to the network. Now, think of what happens when a station wants to connect to a LAN. In the wired world, this is a simple operation. The station uses a cable to plug into an Ethernet jack and it is connected to the network. Even if the network does not explicitly authenticate the station, obtaining physical access to the network provides at least some basic access control if we assume that access to the physical medium is protected. In the wireless world, this *physical-access-authentication* disappears.

For a station to "connect to" or associate with a WLAN, the network-joining operation becomes much more complicated. First, the station must find out which networks it currently has access to. Then, the network must authenticate the station and the station must authenticate the network. Only after this authentication is complete can the station "connect to" or associate with the network (via the AP). Let us go over this process in detail.

---

[3] To be accurate, the shortage of IP V4 addresses. There are more than enough IP V7 addresses available but the deployment of IP V6 has not caught on as fast as its proponents would have liked.

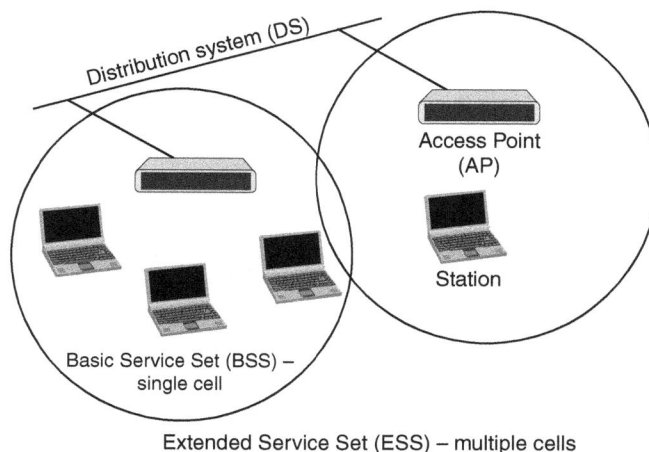

Figure 7.1: 802.11 Network Architecture

Access points in an 802.11 network periodically broadcast beacons. Beacons are management frames that announce the existence of a network. They are used by the APs to allow stations to find and identify a network. Each beacon contains an SSID (*service set identifier*), also called the network name, which uniquely identifies an AP (and therefore a BSA/BSS). When a STA wants to access a network, it has two options: *passive scan* and *active scan* (we will discuss these in detail in the Chapter 8). In the former case, it can scan the channels (i.e., the frequency spectrum) trying to find beacon advertisements from APs in the area. In the latter case, the station sends Probe Requests (either to a particular SSID or with SSID set to 0) over all the channels, one by one. A particular SSID indicates that the station is looking for a particular network. If the concerned AP receives the probe, it responds with a Probe Response. An SSID of 0 indicates that the station is looking to join any network it can access. All APs that receive this Probe Request and that want this particular station to join their network, reply back with a Probe Response. In either case, a station finds out which network(s) it can join.

Next, the station has to choose a network it wishes to join. This decision can be left to the user or the software can make this decision based on signal strengths and other criteria. Once a station has decided that it wants to join a particular network, the authentication process starts. 802.11 provides for two forms of authentication: OSA (*open system authentication*) and SKA (*shared key authentication*). Which authentication is to be used for a particular transaction needs to be agreed upon by both the STA and the network. The STA proposes the authentication scheme it proposes to use in its Authentication Request message. The network may then accept or reject this proposal in its Authentication Response message, depending on how the network administrator has set up the security requirements of the network.

7.4.1 Open System Authentication

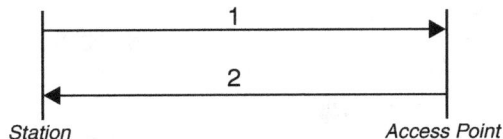

1. Authentication Request : Auth Alg = 0; Trans. Num = 1.
2. Authentication Resp. : Auth Alg = 0; Trans. Num = 2; Status = 0/*

Figure 7.2: Open System Authentication

This is the default authentication algorithm used by 802.11. Here is how it works. Any station that wants to join a network sends an Authentication Request to the appropriate AP. The Authentication Request contains the authentication algorithm that the station wishes to use (0 in case of OSA). The AP replies back with an Authentication Response, thus authenticating the station to join the network[4] if it has been configured to accept OSA as a valid authentication scheme. In other words, the AP does not do any checks on the identity of the station and allows any and all stations to join the network. OSA is exactly what its name suggests: Open System Authentication. The AP (network) allows any station (that wishes to join) to join the network. Using OSA, therefore, means using no authentication at all.

It is important to note here that the AP can enforce the use of authentication. If a station sends an Authentication Request requesting to use OSA, the AP may deny the station access to the network if the AP is configured to enforce SKA on all stations.

7.4.2 Shared Key Authentication

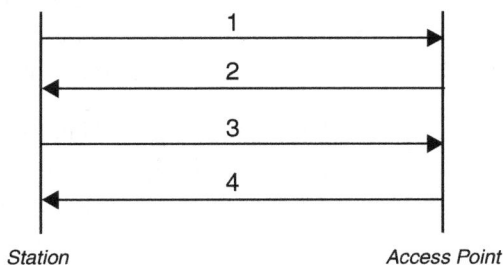

1. Authentication Request : Auth Alg = 1; Trans. Num = 1.
2. Authentication Resp : Auth Alg = 1; Trans. Num = 2; Data = 128-byte random number.
3. Authentication Resp : Auth Alg = 1; Trans. Num = 3; Data = Encrypted (128-byte number rcvd in.
4. Authentication Resp : Auth Alg = 1; Trans. Num = 4; Status = 0/*.

Figure 7.3: Shared Key Authentication

---

[4] The Authentication Request from the station may be denied by the AP for reasons other than Authentication Failure, in which case the status field will be non-zero.

SKA is based on the challenge-response system. SKA divides stations into two groups. Group One consists of stations that are allowed access to the network, and Group Two consists of all other stations. Stations in Group One share a secret key, which stations in Group Two don't know. By using SKA, we can ensure that only stations belonging to Group One are allowed to join the network.

Using SKA requires 1) that the station and the AP be capable of using WEP, and 2) that the station and the AP have a preshared key. The second requirement means that a shared key must be distributed to all stations that are allowed to join the network before attempting authentication. How this is done is not specified in the 802.11 standard. Figure 7.3 above explains how SKA works in detail.

When a station wants to join a network, it sends an Authentication Request to the appropriate AP which contains the authentication algorithm it wishes to use (1 in case of SKA). On receiving this request, the AP sends an Authentication Response back to the station. This Authentication Response contains a challenge text. The challenge text is a 128-byte number generated by the *pseudo-random-number-generator* (also used in WEP) using the preshared secret key and a random *initialization vector* (IV). When the station receives this random number (the challenge), it encrypts the random number using WEP[5] and its own IV to generate a response to the challenge. Note that the IV that the station uses for encrypting the challenge is different from (and independent of) the IV that the AP used for generating the random number. After encrypting the challenge, the station sends the encrypted challenge and the IV it used for encryption back to the AP as the response to the challenge. On receiving the response, the AP decrypts the response using the preshared keys and the IV that it receives as part of the response. The AP compares the decrypted message with the challenge it sent to the station. If these are the same, the AP concludes that the station wishing to join the network is one of the stations that knows the secret key, and therefore the AP authenticates the station to join the network.

The SKA mechanism allows an AP to verify that a station is one of a select group of stations. The AP verifies this by ensuring that the station knows a secret, which is the preshared key. If a station does not know the key, it will not be able to respond "correctly" to the challenge. Thus, the strength of SKA lies in keeping the shared key a secret.

---

[5] WEP is described in section 7.2.1.

7.4.3 Authentication and Handoffs

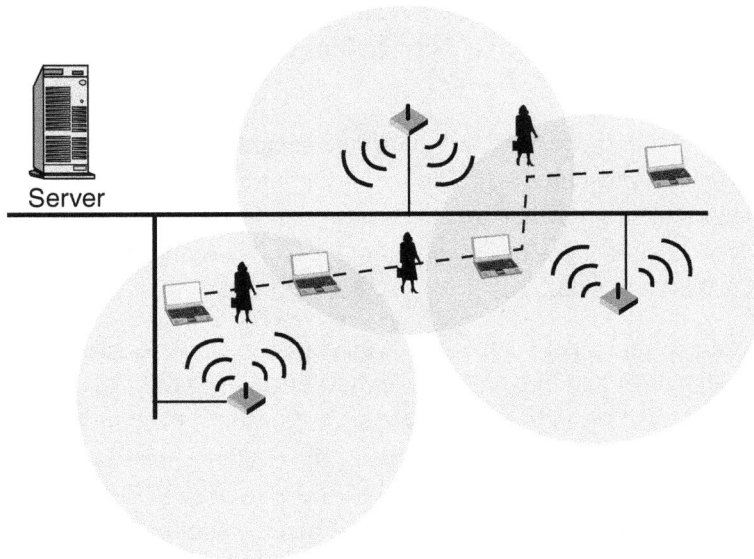

Figure 7.4: Inter-BSS/Intra-ESS Handoffs

If a station is mobile while accessing the network, it may leave the range of one AP and enter into the range of another AP, as shown in Figure 7.4. In this section we see how authentication fits in with mobility.

A STA may move inside a BSA (intra-BSA), between two BSAs (inter-BSA) or between two ESAs (inter-ESAs). In the intra-BSA case, the STA is static for all handoff purposes. Inter-ESA roaming requires support from higher layers (MobileIP, for example) since ESAs communicate with each other at layer 3.

As we will discuss in the next chapter, 802.11 has provisions for inter-BSA roaming. In its simplest form, a STA capable of mobility keeps track of the *received signal strength* (RSS) of the beacon with which it is associated. When this RSS value falls below a certain threshold, the STA starts to scan for stronger beacon signals available to it, using either active or passive scanning. This procedure continues until the RSS of the current beacon returns above the threshold (in which case, the STA stops scanning for alternate beacons) or until the RSS of the current beacon falls below the break-off threshold, in which case the STA decides to hand off to the strongest beacon available. When this situation is received, the STA disconnects from its prior AP and connects to the new AP afresh (just as if it had switched on in the BSA of the new AP). In fact, the association with the prior AP is not "carried over"/"handed off" transparently to the new AP: the STA disconnects with the old AP and then connects with the new AP.

To connect to the new AP, the STA starts the connection procedure afresh. This means that the process of associating (and authenticating) to the new AP is the same as it is for a STA that has just powered on in this BSS. In other words, the prior AP and the post AP do not coordinate among themselves to achieve a handoff.[6] Analysis[7] has shown that authentication delays are the second biggest contributors to handoff times next only to channel scanning/probing time. This reauthentication delay becomes even more of a bottleneck for real-time applications like voice. Although, this is not exactly a security loophole, it is a drawback of using the security. We will discuss this in detail in Chapter 8.

7.4.4 What's Wrong with 802.11 Authentication?

Authentication mechanisms suggested by 802.11 suffer from many drawbacks. As we saw, 802.11 specifies two modes of authentication—OSA and SKA. OSA provides no authentication and is irrelevant here.

SKA works on a challenge-response system as explained in section 7.4.2. The AP expects that the challenge it sends to the STA will be encrypted using an IV and the *preshared key*. As described in section 7.2.1, there is no method specified in WEP for each STA to be assigned a unique key. Instead all STAs and the AP are configured with the same key. This means that, even when an AP authenticates a STA using the SKA mode, all it ensures is that the STA belongs to a group of STAs that know the preshared key. There is no way for the AP to determine the exact identity of the STA that is trying to authenticate to the network and access it.[8]

To make matters worse, many 802.11 deployments share keys across access points. This increases the size of the group to which a STA can be traced. All STAs sharing a single preshared secret key also makes it very difficult to remove a STA from the allowed set of STAs, since this would involve changing (and redistributing) the shared secret key to all stations.

There is another issue with 802.11 authentication: it is one-way. Even though it provides a mechanism for the AP to authenticate the STA, it has no provision for the STA to be able to authenticate the network. This means that a rogue AP may be able to hijack the STA by establishing a session with it. This is a very plausible scenario, given the plummeting cost of APs. Since, the STA can never find out that it is communicating with a rogue AP, the rogue AP virtually has access to everything that the STA sends to it.

Finally, SKA is based on WEP. It therefore suffers from all the drawbacks that WEP suffers from too (see section 7.5).

---

[6] To be accurate, the base IEEE 802.11 standard does not specify how the two APs should communicate with each other. There do exist proprietary solutions by various vendors, as well as on-going standard work which enables inter-AP communication to improve handoff performance. The next chapter discusses some of these extensions.

[7] *An Empirical Analysis of the IEEE 802.11 MAC layer Handoff Process*, Mishra, etc.

[8] MAC addresses can be used for this purpose but they are not cryptographically protected, in that it is easy to spoof MAC addresses.

7.5 Confidentiality in 802.11

WEP uses a pre-established/preshared set of keys. Figure 7.5 shows how WEP is used to encrypt an 802.11 MPDU (*Media-access control Protocol Data Unit*). Note that Layer 3 (usually IP) hands over an MSDU (*Media-access control Service Data Unit*) to the 802.11 MAC layer. The 802.11 protocol may then fragment the MSDU into multiple MPDUs if it is required to use the channel efficiently.

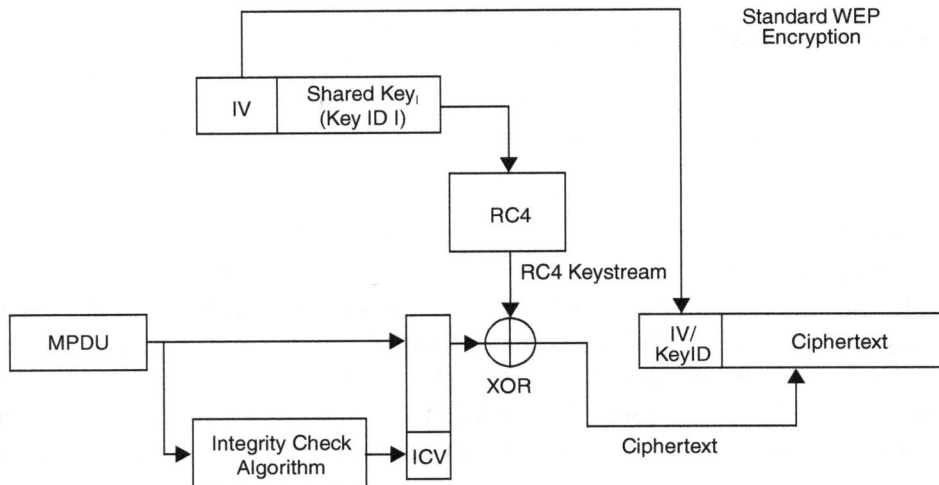

Figure 7.5: Standard WEP Encryption

The WEP process can be broken down into the following steps:

Step 1: Calculate the ICV (*integrity check value*) over the length of the MPDU and append this 4-byte value to the end of the MPDU. Note that ICV is another name for MIC (*message integrity check*). We see how this ICV value is generated in section 7.6.

Step 2: Select a master key to be used from one of the four possible preshared secret keys. See section 7.2.1 for the explanation of the four possible preshared secret keys.

Step 3: Select an IV and concatenate it with the master key to obtain a key stream. WEP does not specify how to select the IV. The IV selection process is left to the implementation.

Step 4: The key generated in Step 3 is then fed to an RC4 key generator. The resulting RC4 key stream is then XORed with the MPDU + ICV generated in Step 1 to generate the cipher text.

Step 5: A 4 byte header is then appended to the encrypted packet that contains the 3-byte IV value and a 1-byte key-ID specifying which one of the four preshared secret keys is being used as the master key.

The WEP process is now completed. An 802.11 header is then appended to this packet and it is ready for transmission. The format of this packet is shown in Figure 7.6.

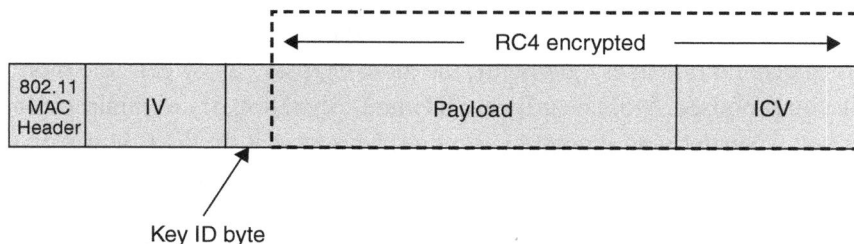

Figure 7.6: Format of WEP Encrypted 802.11 Packet

7.5.1 What's Wrong with WEP?

WEP uses RC4 (a stream cipher) in synchronous mode for encrypting data packets. Synchronous stream ciphers require that the key generators at the two communicating nodes must be kept synchronized (by some external means) because the loss of a single bit of a data stream encrypted under the cipher causes the loss of ALL data following the lost bit. In brief, this is so because data loss desynchronizes the key stream generators at the two endpoints. Since data loss is widespread in the wireless medium, a synchronous stream cipher is not the right choice. This is one of the most fundamental problems of WEP. It uses a cipher not suitable for the environment it operates in.

It is important to re-emphasize that the problem here is not the RC4 algorithm.[9] The problem is that a stream cipher is not suitable for a wireless medium where packet loss is widespread. SSL (secure sockets layer) uses RC4 at the application layer successfully because SSL (and therefore RC4) operates over a reliable data channel that does not lose any data packets and can therefore guarantee perfect synchronization between the two endpoints.

The WEP designers were aware of the problem of using RC4 in a wireless environment. They realized that, due to the widespread data loss in the wireless medium, using a synchronous stream cipher across 802.11 frame boundaries was not a viable option. As a solution, WEP attempted to solve the synchronization problem of stream ciphers by shifting the synchronization requirement from a session to a packet. In other words, since the synchronization between the endpoints is not perfect (and subject to packet loss), 802.11 changes keys for every packet. This way, each packet can be encrypted/decrypted irrespective of the previous packets loss. Compare this with SSL's use of RC4, which can afford to use a single key for a complete session. In effect, since the wireless medium is prone to data loss, WEP has to use

---

[9] Though loopholes in the RC4 algorithm have been discovered.

a single packet as the "synchronization unit" rather than a complete session. This means that WEP uses a unique key for each packet.

Using a separate key for each packet solves the synchronization problem but introduces problems of its own. Recall that, to create a per-packet key, the IV is simply concatenated with the master key. As a general rule in cryptography, the more exposure a key gets, the more it is susceptible to be compromised. Most security architectures, therefore, try to minimize the exposure of the master key when deriving secondary (session) keys from it. In WEP, however, the derivation of the secondary (per-packet) key from the master key is too trivial to hide the master key.

Another aspect of WEP security is that the IV that is concatenated with the master key to create the per-packet key is transmitted in clear-text with the packet too. Since the IV is transmitted in the clear with each packet, an eavesdropper already has access to the first 3 bytes of the per-packet key.

The above two weaknesses make WEP susceptible to an FMS (*Fluhrer-Mantin-Shamir*) attack which uses the fact that simply concatenating the IV (available in plain text) to the master key leads to the generation of a class of RC4 weak keys. The FMS attack exploits the fact that the WEP creates the per-packet key by simply concatenating the IV with the master key. Since the first 24 bits of each per-packet key is the IV (which is available in plain text to an eavesdropper),[10] the probability of using "weak-keys"[11] is very high. Note that the FMS attack is a weakness in the RC4 algorithm itself. However, it is the way that the per-packet keys are constructed in WEP that makes the FMS attack much more effective in 802.11 networks.

The FMS attack relies on the ability of the attacker to collect multiple 802.11 packets that have been encrypted with weak keys. Limited key space (leading to key reuse) and availability of IV in clear-text, which forms the first 3-bytes of the key, makes the FMS attack a very real threat in WEP. This attack is made even more potent in 802.11 networks by the fact that the first 8 bytes of the encrypted data in every packet are known to be the SNAP header. This means that simply XORing the first 2 bytes of the encrypted pay-load with the well-known SNAP header yields the first 2 bytes of the generated key-stream. In the FMS attack, if the first 2 bytes of enough key-streams are known, then the RC4 key can be recovered. Thus, WEP is an ideal candidate for an FMS attack.

The FMS attack is a very effective attack but is by no means the only attack that can exploit WEP weaknesses. Another such attack stems from the fact that one of the most important requirements of a synchronous stream cipher (like RC4) is that the same key should not be re-used *EVER*. We look at what 802.11 needs to achieve this. Since we need a new key for every

---

[10] Remember that each WEP packet carries the IV in plain-text format prepended to the encrypted packet.

[11] Use of certain key values leads to a situation where the first few bytes of the output are not all that random. Such keys are known as weak-keys. The simplest example is a key value of 0.

single packet to make the network really secure, 802.11 needs a very large key space—i.e., a large number of unique keys. The number of unique keys available is a function of the key length. What is the key length used in WEP? Theoretically it is 74 bits. The devil, however, is in the details. How is the 74-bit key constructed? 24 bits come from the IV and 40 bits come from the base-key. Since the 40-bit master key never changes in most 802.11 deployments,[12] we must ensure that we use different IVs for each packet in order to avoid key reuse. Since the master key is fixed in length and the IV is only 24 bits long, the effective key-length of WEP is 24 bits. Therefore, the key space for the RC4 is 2^N where N is the length of the IV. 802.11 specified the IV length as 24.

To put things in perspective, realize that if we have a 24-bit IV (2^{24} keys in the key-space), a busy base station which is sending 1500 byte-packets @ 11 Mbps will exhaust all keys in the key space in $(1500*8 * 2^{24})/(11*10^7)$ seconds or approximately 5 hours. On the other hand, RC4 in SSL would use the same key space for 2^{24} (=10^7) sessions. Even if the application has 10,000 sessions per day, the key space would last for 3 years. In other words, an 802.11 BS using RC4 has to reuse the same key in approximately 5 hours, whereas an application using SSL RC4 can avoid key reuse for approximately 3 years. This shows clearly that the fault lies not in the cipher but in the way it is being used. Going beyond an example, analyses of WEP [3], [4] have shown that there is a 50% chance of key reuse after 4823 packets and there is a 99% chance of collision after 12,430 packets. These are dangerous numbers for a cryptographic algorithm.

Believe it or not, it gets worse. 802.11 specifies no rules for IV selection. This in turn means that changing the IV with each packet is optional. This effectively means that 802.11 implementations may use the same key to encrypt all packets without violating the 802.11 specifications. Most implementations, however, vary from randomly generating the IV on a per-packet basis to using a counter for IV generation. WEP does specify that the IV be changed "frequently." Since this is vague, it means that an implementation that generates per-packet keys (more precisely the per-MPDU key) is 802.11-compliant, and so is an implementation which reuses the same key across MPDUs.

7.6 Data Integrity in 802.11

To ensure that a packet has not been modified in transit, 802.11 uses an ICV (*integrity check value*) field in the packet. ICV is another name for MIC (*message integrity check*). An ICV is a small tag that is appended to a packet. This tag is generated in a cryptographic fashion (for example, a one-way hash of the packet contents and a key). The idea behind the ICV/MIC is that the receiver should be able to detect data modifications or forgeries by calculating the ICV over the received data and comparing it with the ICV attached in the message. Figure 7.7 shows the complete picture of how WEP and CRC32 work together to create the MPDU for transmission.

---

[12] This weakness stems from the lack of a key-establishment/key-distribution protocol in WEP.

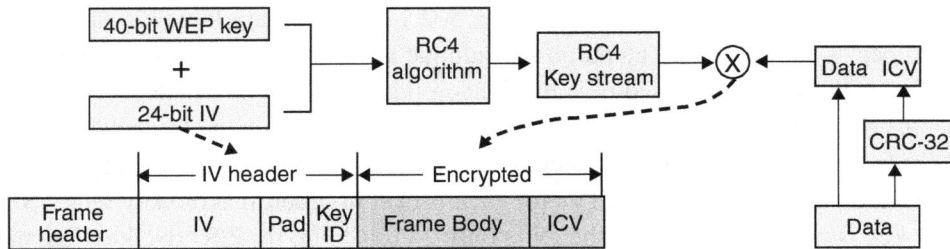

Figure 7.7: Security in Base 802.11

The underlying assumption is that, if Eve modifies the data in transit, she should not be able to modify the ICV appropriately to force the receiver into accepting the packet. In WEP, ICV is implemented as a CRC-32 (*cyclic redundancy check-32 bits*) checksum, which breaks this assumption. The reason for this is that CRC-32 is linear and is not cryptographically computed. This means that the CRC32 has the following interesting property:

$$CRC(X \oplus Y) = CRC(X) \oplus CRC(Y)$$

Now, if X represents the payload of the 802.11 packet over which the ICV is calculated, the ICV is $CRC(X)$, which is appended to the packet. Consider an intruder who wishes to change the value of X to Z. To do this, he calculates $Y = X \oplus Z$. Then he captures the packet from the air-interface, XORs X with Y and then XORs the ICV with $CRC(Y)$. Therefore, the packet changes from $\{X, CRC(X)\}$ to $\{X \oplus Y, CRC(X) \oplus CRC(Y)\}$ or simply $\{X \oplus Y, CRC(X \oplus Y)\}$. If the intruder now retransmits the packets to the receiver, the receiver would have no way of telling that the packet was modified in transit. This means that we can change bits in the payload of the packet while preserving the integrity of the packet, if we also change the corresponding bits in the ICV of the packet.

Note that an attack like the one described above works because flipping bit x in the message results in a deterministic set of bits in the CRC that must be flipped to produce the correct checksum of the modified message. This property stems from the linearity of the CRC32 algorithm.

Realize that, even though the ICV is encrypted (cryptographically protected) along with the rest of the payload in the packet, it is not cryptographically computed—i.e., calculating the ICV does not involve keys and cryptographic operations. Simply encrypting the ICV does not prevent an attack like the one discussed above. This is so because the flipping of a bit in the cipher text carries through after the RC4 decryption into the plain text because:

$RC4(k, X \oplus Y) = RC4(k, X) \oplus Y$ and therefore:

$$RC4(k, CRC(X \oplus Y)) = RC4(k, CRC(X)) \oplus CRC(Y)$$

The problem with the message integrity mechanism specified in 802.11 is not only that it uses a linear integrity check algorithm (CRC32) but also the fact that the ICV does not protect all the information that needs to be protected from modification. Recall from section 7.5 that the ICV is calculated over the MPDU data; that is, the 802.11 header is not protected by the ICV. This opens the door to redirection attacks as explained in the following paragraphs.

Consider an 802.11 BSS where an 802.11 STA (Alice) is communicating with a wired station (Bob). Since the wireless link between Alice and the AP (access point) is protected by WEP and the wired link between Bob and access point is not,[13] it is the responsibility of the AP to decrypt the WEP packets and forward them to Bob. Now, Eve captures the packets being sent from Alice to Bob over the wireless link. She then modifies the destination address to another node, say C (Charlie), in the 802.11 header and retransmits them to the AP. Since the AP does not know any better, it decrypts the packet and forwards it to Charlie. Eve, therefore, has the AP decrypt the packets and forward them to a destination address of choice.

The simplicity of this attack makes it extremely attractive. All Eve needs is a wired station connected to the AP and she can eavesdrop on the communication between Alice and Bob without needing to decrypt any packets herself. In effect, Eve uses the infrastructure itself to decrypt any packets sent from an 802.11 STA via an AP. Note that this attack does not necessarily require that one of the communicating stations be a wired station. Either Bob or Charlie (or both) could just as easily be other 802.11 STAs that do not use WEP. The attack would still hold since the responsibility of decryption would still be with the AP. The bottom line is that the redirection attack is possible because the ICV is not calculated over the 802.11 header. There is an interesting security lesson here. A system can't have confidentiality without integrity, since an attacker can use the redirection attack and exploit the infrastructure to decrypt the encrypted traffic.

Another problem stemming from the weak integrity protection in WEP is the threat of a replay attack. A replay attack works by capturing 802.11 packets transmitted over the wireless interface and then replaying (retransmitting) the captured packet(s) later on with (or without) modification, such that the receiving station has no way to tell that the packet it is receiving is an old (replayed) packet. To see how this attack can be exploited, consider a hypothetical scenario where Alice is an account holder, Bob is a bank and Eve is another account holder in the bank. Suppose Alice and Eve do some business and Alice needs to pay Eve $500. So, Alice connects to Bob over the network and transfers $500 from her account to Eve. Eve, however, is greedy. She knows Alice is going to transfer money, so she captures all data going from Alice to Bob. Even though Eve does not know what the messages say, she has a pretty good guess that these messages instruct Bob to transfer $500 from Alice's account to Eve's. So,

---

[13] WEP is an 802.11 standard used only on the wireless link.

Eve waits a couple of days and replays these captured messages to Bob. This may have the effect of transferring another $500 from Alice's account to Eve's account unless Bob has some mechanism for determining that it is being replayed the messages from a previous session.

Replay attacks are usually prevented by linking the integrity protection mechanism to either timestamps and/or session sequence numbers. However WEP does not provide for any such protection.

7.7 Loopholes in 802.11 Security

To summarize, here is the list of things that is wrong with 802.11 security:

a. 802.11 does not provide any mechanism for key establishment over an insecure medium. This means key sharing among STAs in a BSS and sometimes across BSSs.

b. WEP uses a synchronous stream cipher over a medium, where it is difficult to ensure synchronization during a complete session.

c. To solve the previous problem, WEP uses a per-packet key by concatenating the IV directly to the preshared key to produce a key for RC4. This exposes the base-key / master-key to attacks like FMS.

d. Since the master-key is usually manually configured and static and since the IV used in 802.11 is just 24 bits long, this results in a very limited key-space.

e. 802.11 specifies that changing the IV with each packet is optional, thus making key reuse highly probable.

f. CRC-32 used for message integrity is linear.

g. The ICV does not protect the integrity of the 802.11 header thus opening the door to redirection attacks.

h. There is no protection against replay attacks.

i. No support for a STA to authenticate the network.

Note that the limited size of the IV figures much lower in the list than one would expect. This emphasizes the fact that simply increasing the IV size would not improve WEP's security considerably. The deficiency of the WEP encapsulation design arises from attempts to adapt RC4 to an environment for which it is poorly suited.

7.8 WPA

When the loopholes in WEP, the original 802.11 security standard, had been exposed, IEEE formed a Task Group: 802.11i with the aim of improving upon the security of 802.11 networks. This group came up with the proposal of a RSN (*robust security network*). An RSN is an 802.11 network that implements the security proposals specified by the 802.11i group and allows only RSN-capable devices to join the network, thus allowing no "holes." The term "hole" is used to refer to a non-802.11i compliant STA which, by virtue of not following the 802.11i security standard, could make the whole network susceptible to a variety of attacks.

Since making a transition from an existing 802.11 network to a RSN cannot always be a single-step process (we will see why in a moment), 802.11i allows for a TSN (*transitional security network*) which provides for the existence of both RSN and WEP nodes in an 802.11 network. As the name suggests, this kind of a network is specified only as a transition point and all 802.11 networks are finally expected to move to a RSN. The terms RSN and 802.11i are sometimes used interchangeably to refer to this security specification.

The security proposal specified by Task-Group-i uses the AES (*advanced encryption standard*) in its default mode. One obstacle in using AES is that it is not backward compatible with existing WEP hardware, because AES requires the existence of a new, more powerful hardware engine. This means that there is also a need for a security solution that can operate on existing hardware. This was a pressing need for vendors of 802.11 equipment and is where the *Wi-Fi Alliance* came into the picture.

The Wi-Fi Alliance is an alliance of major 802.11 vendors formed with the aim of ensuring product interoperability. To improve the security of 802.11 networks without requiring a hardware upgrade, the Wi-Fi Alliance adopted TKIP (*Temporal Key Integrity Protocol*) as the security standard that needs to be deployed for Wi-Fi certification. This form of security has therefore come to be known as WPA (*Wi-Fi protected access*). WPA is basically a pre-standard subset of 802.11i that includes the key management and the authentication architecture (802.1X) specified in 802.11i. The biggest difference between WPA and 802l.11i (which has also come to be known as WPA2) is that, instead of using AES for providing confidentiality and integrity, WPA uses TKIP for encryption and an algorithm called MICHAEL for per-packet authentication. We look at TKIP/WPA in this section and the 802.11i/WPA2 using AES in the next section.

TKIP was designed to fix WEP loopholes while operating within the constraints of existing 802.11 equipment (APs, WLAN cards, etc.). To understand what we mean by the "constraints of existing 802.11 hardware," we need to dig a little deeper. Most 802.11 equipment consists of some sort of a WLAN-NIC (aka WLAN-adapter), which enables access to an 802.11 network. A WLAN-NIC usually consists of a small microprocessor, some firmware, a small amount of memory and a special-purpose hardware engine. This hardware engine is dedicated to WEP implementation since software implementations of WEP are too slow. To be precise, the WEP encryption process is implemented in hardware. The hardware encryption takes the IV, the base (or master) key and the plain-text data as the input and produces the encrypted output (cipher-text). One of the most severe constraints for TKIP designers was that the hardware engine cannot be changed. We see in this section how WEP loopholes were closed, given these constraints.

7.8.1 Key Establishment

One of the biggest WEP loopholes is that it specifies no key-establishment protocol and relies on the concept of preshared secret keys to be established using some out-of-band mechanism. Realize that this is a system-architecture problem. In other words, solving this problem requires support from multiple components (the AP, the STA and usually also a back-end authentication server) in the architecture.

One of the important realizations of the IEEE 802.11i task group was that 802.11 networks were being used in two distinct environments: the home network and the enterprise network. These two environments had distinct security requirements and different infrastructure capacities to provide security. Therefore, 802.11i specified two distinct security architectures. For the enterprise network, 802.11i specifies the use of IEEE 802.1X for key establishment and authentication. As we will see in our discussion in the next section, 802.1X requires the use of a back-end authentication server. Deploying a back-end authentication server is not usually feasible in a home environment. Therefore, for home deployments of 802.11, 802.11i allows the use of "out-of-band mechanism" (read manual configuration) for key establishment.

We look at the 802.1X architecture in the next section and see how it results in the establishment of a master key (MK). In this section, we assume that the two communicating endpoints (the STA and the AP) already share an MK that has either been configured manually at the two endpoints (WEP architecture) or has been established using the authentication process (802.1X architecture). This section looks at how this MK is used in WPA.

Recall that a major loophole in WEP was the manner[14] in which this master key was used, which made it vulnerable to compromise. WPA solves this problem by reducing the exposure of the master key, thus making it difficult for an attacker to discover the master key. To achieve this, WPA adds an additional layer to the key hierarchy used in WEP. Recall from section 7.5 that WEP uses the master key for authentication and to calculate the per-packet key. In effect, there is a two-tier key hierarchy in WEP: the master (preshared secret) key and the per-packet key.

---

[14] The per-packet key is obtained by simply concatenating the IV with the preshared secret key. Therefore, a compromised per-packet key exposes the preshared secret key.

WEP
WPA
(used with 802.1X)
WPA
(used without 802.1X)

Master-Secret
(used by authentication process; certificate; password, etc.)

Master-Secret
(User Password)

By-product of 802.1X-based authentication process

Can be specified by Network Administrator

Master-Key
(Pre-Shared / Manually Configured)
40 bits / 104 bits

PMK
(Pair-wise Master Key)
256 bits

PRF-512 (PMK, "Pair-wise Key Expansion", MACf || MAC2 || Noncef || Nonce2)

PTK (Pair-wise Transient Keys)

| Data Encryption-Key 128 bits | Data MIC-Key 128 bits | EAPoL Encryption-Key 128 bits | EAPoL MIC-Key 128 bits |

Prepend with IV

Phase-1 and Phase-2 Key Mixing

Per-Packet-Encryption-Key

Per-Packet-Encryption-Key

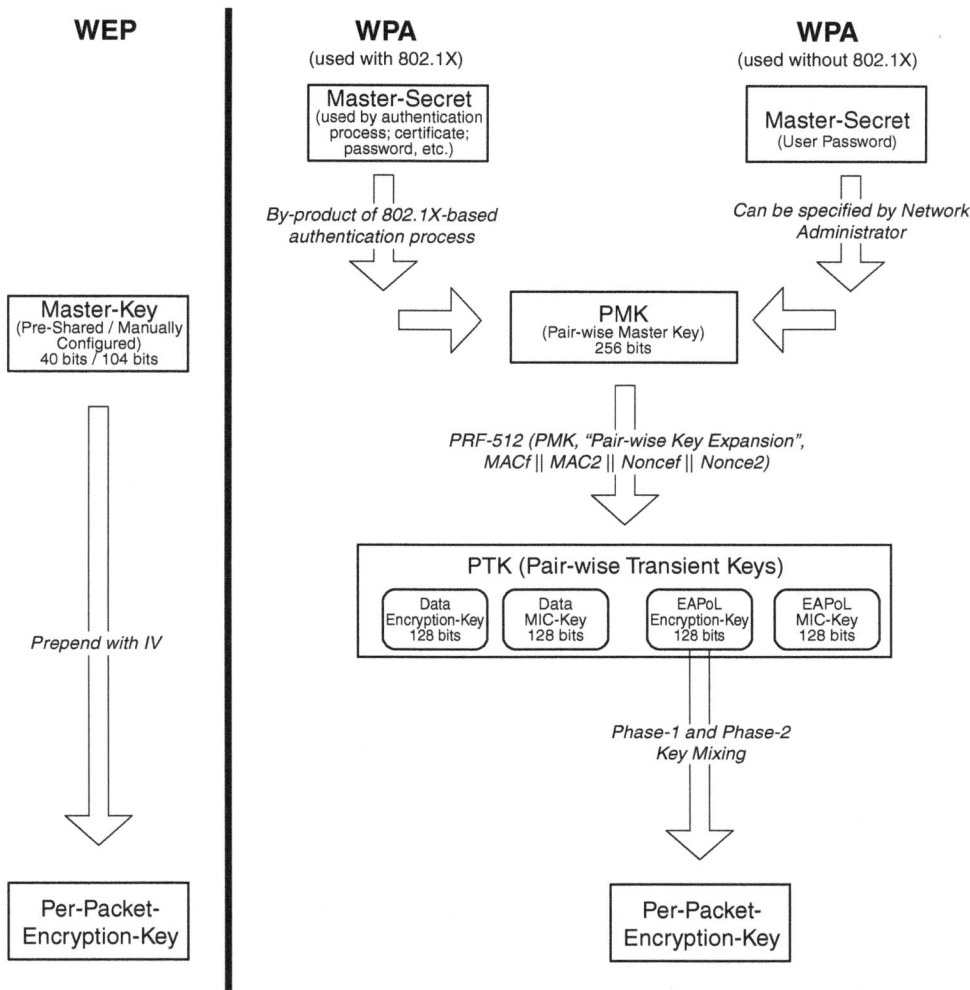

Figure 7.8: Key Generation in WEP, WPA and WPA2

WPA extends the two-tier key hierarchy of WEP to a multitier hierarchy (see Figure 7.8). At the top level is still the master key, referred to as the PMK (*pair-wise master key*) in WPA. The next level in the key hierarchy is the PTK (*pair-wise transient key*), derived from the PMK. The final level is the per-packet keys, which are generated by feeding the PTK to a key-mixing function. Compared with the two-tier WEP key hierarchy, the three-tier key hierarchy of WPA avoids exposing the PMK in each packet by introducing the concept of PTK.

As we saw, WPA does not specify how the master key (PMK in WPA) is established. The PMK, therefore, may be a preshared[15] secret key (WEP-design) or a key derived from an authentication process like 802.1X.[16] WPA does require that the PMK be 257 bits (or 32 bytes) long. Since a 32-byte key is too long for humans to remember, 802.11 deployments using preshared keys may allow users to enter a shorter password, which may then be used as a seed to generate the 32-byte key.

The next level in the key hierarchy is the PTK, which are basically session keys. The term PTK is used to refer to a set of session keys that consists of four keys, each of which is 128 bits long. These four keys are as follows: an encryption key for data, an integrity key for data, an encryption key for Extensible Authentication Protocol over LAN (EAPoL) messages and an integrity key for EAPoL messages. Note that the term "session" here refers to the association between a STA and an AP. Every time a STA associates with an AP, it is the beginning of a new session and this results in the generation of a new PTK (set of keys) from the PMK. Since the session keys are valid only for a certain period of time, they are also referred to as *temporal keys*. The PTK are derived from the PMK using a PRF (pseudo random function). The PRFs used for derivation of the PTK (and *Nonces*, defined below) are explicitly specified by WPA and are based on the HMAC-SHA algorithm:

PTK = PRF–512(PMK, "Pairwise key expansion", AP_MAC || STA_MAC || ANonce || SNonce)

To obtain the PTK from the PMK we need five input values: the PMK, the MAC addresses of the two endpoints involved in the session and one Nonce each from the two endpoints. The use of the MAC addresses in the derivation of the PTK ensures that the keys are "bound" to sessions between the two endpoints and increases the effective key space of the overall system.

Since we want to generate a different set of session keys from the same PMK for each new session,[17] we need to add another input into the key-generation mechanism that changes with each session. This input is the *Nonce*. The concept of Nonce is best understood by realizing that it is short for "number-once." The value of Nonce is thus arbitrary, except that a Nonce value is never used again[18]—it is basically a number which is used only once. In our context, a Nonce is a unique number (generated randomly) that can distinguish between two sessions established between a given STA and an AP at different points in time. The two Nonces involved in PTK generation are generated one each by the two endpoints involved in the

---

[15] As we saw, this usually means that the keys are manually configured.

[16] It is expected that most enterprise deployments of 802.11 would use 802.1X while the preshared secret key method (read manual configuration) would be used by residential users.

[17] If a STA disconnects from the AP and connects back with an AP at a later time, these are considered two different sessions.

[18] To be completely accurate, Nonce values are generated so that the probability of the same value being generated twice is very low.

session—i.e., the STA (SNonce) and the AP (ANonce). WPA specifies that a Nonce should be generated as follows:

$$\text{ANonce} = \text{PRF–257}(\text{Random Number, "Init Counter", AP\_MAC} \parallel \text{Time})$$

$$\text{SNonce} = \text{PRF–257}(\text{Random Number, "Init Counter", STA\_MAC} \parallel \text{Time})$$

The important thing to note is that the PTK are effectively shared between the STA and the AP and are used by both the STA and the AP to protect the data/EAPoL messages they transmit. Therefore the input values required for derivation of PTK from the PMK come from *both* the STA and the AP. Note also that the key-derivation process can be executed in parallel at both endpoints of the session (the STA and the AP) once the Nonces and the MAC addresses have been exchanged. Thus, both the STA and the AP can derive the same PTK from the PMK simultaneously.

The next step in the key-hierarchy tree is to derive per-packet keys from the PTK. WPA improves also upon this process significantly. Recall from section 7.5 that the per-packet key was obtained by simply concatenating the IV with the master key in WEP. Instead of simply concatenating the IV with the master key, WPA uses the process shown in Figure 7.9 to obtain the per-packet key. This process is known as *per-packet key mixing*.

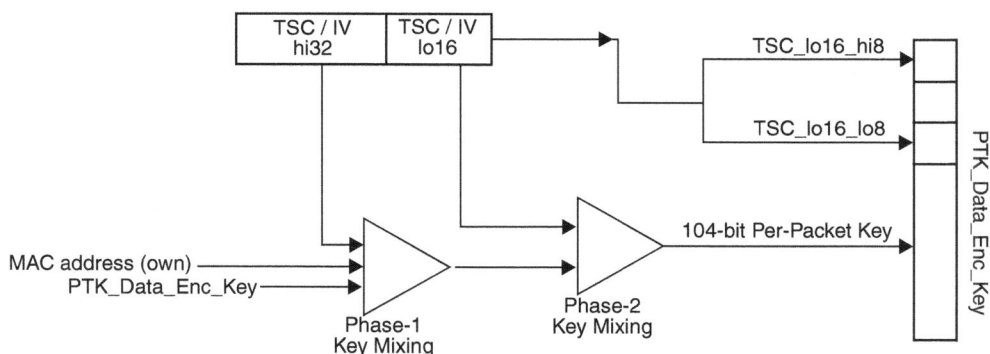

Figure 7.9: Per-Packet Key Generation

In phase one, the session data encryption key is combined with the high-order 32 bits of the IV and the MAC address. The output from this phase is combined with the lower-order 17 bits of the IV and fed to phase two, which generates the 104-bit per-packet key. There are many important features to note in this process:

a. It assumes the use of a 48-bit IV (more of this in section 7.8.2)

b. The size of the encryption key is still 104 bits, thus making it compatible with existing WEP hardware accelerators.

c. Since generating a per-packet key involves a hash operation, which is computation intensive for the small MAC processor in existing WEP hardware, the process is split into two phases. The processing-intensive part is done in phase one, whereas phase two is much less computation intensive.

d. Since phase one involves the high-order 32 bits of the IV, it needs to be done only when one of these bits change—i.e., once in every 75,537 packets.

e. The key-mixing function makes it very hard for an eavesdropper to correlate the IV and the per-packet key used to encrypt the packet.

7.8.2 Authentication

As we said in the previous section, 802.11i specified two distinct security architectures. For the home network, 802.11i allows the manual configuration of keys just like WEP. For the enterprise network, however, 802.11i specifies the use of IEEE 802.1X for key establishment and authentication.

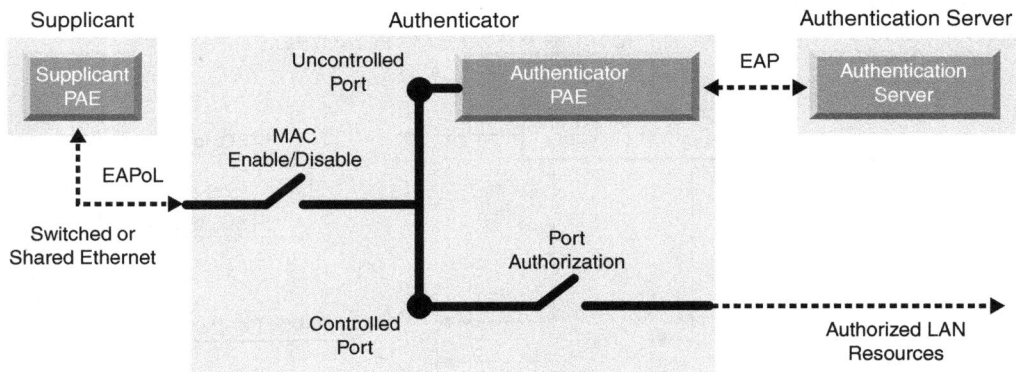

Figure 7.10: Authentication Architecture

802.1X is closely architected along the lines of EAPoL. Figure 7.10 shows the conceptual architecture of EAPoL and Figure 7.11 shows the overall system architecture of EAPoL. The controlled port is open only when the device connected to the port has been authorized by 802.1x. On the other hand, the uncontrolled port provides a path for EAPOL traffic ONLY. Figure 7.10 shows how access to even the uncontrolled port can be limited using MAC filtering.[19] This scheme is sometimes used to deter DoS attacks.

---

[19] Allowing only STAs which have a MAC address which is "registered" or "known" to the network.

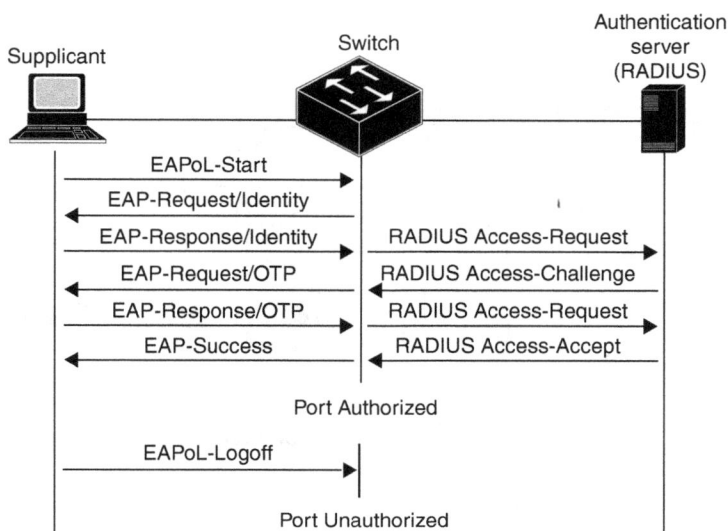

Figure 7.11: EAPOL

EAP specifies three network elements: the supplicant, the authenticator and the authentication server. For EAPoL, the end user is the supplicant, the Layer-2 (usually Ethernet) switch is the authenticator controlling access to the network using logical ports, and the access decisions are taken by the backend authentication server after carrying out the authentication process. Which authentication process to use (MD5, TLS etc.) is for the network administrator to decide.

EAPoL can be easily adapted to be used in the 802.11 environment, as shown in Figure 7.12. The STA is the supplicant, the AP is the authenticator controlling access to the network and there is a backend authentication server. The analogy is all the more striking if you consider that an AP is, in fact, just a layer-2 switch, with a wireless and a wired interface.

There is, however, one interesting piece of detail that needs attention. The 802.1X architecture carries the authentication process between the supplicant (STA) and the backend authentication server.[20] This means that the master key (resulting from an authentication process like TLS) is established between the STA and backend server. However, confidentiality and integrity mechanisms in the 802.11 security architecture are implemented between the AP and the STA. This means that the session (PTK) and per-packet keys (which are derived from the PMK) are needed at the STA and the AP. The STA already has the PMK and can derive the PTK and the per-packet keys. However, the AP does not yet have the PMK. Therefore, what is needed is a mechanism to get the PMK from the authentication server to the AP securely.

---

[20] With the AP controlling access to the network using logical ports.

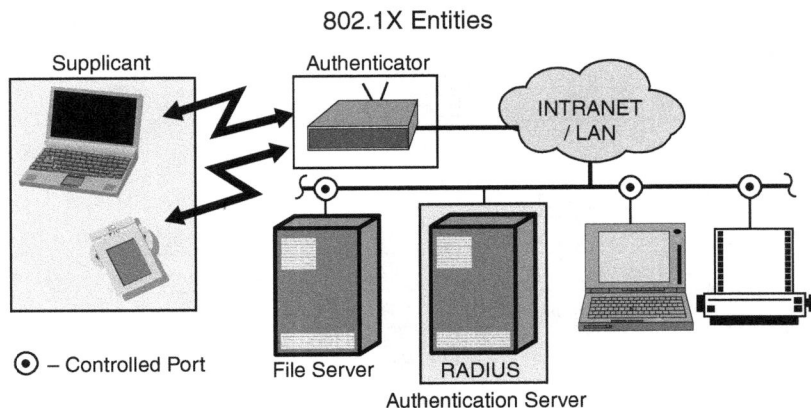

Figure 7.12: Authentication Overview

Recall that in the 802.1X architecture, the result of the authentication process is conveyed by the authentication server to the AP so that the AP may allow or disallow the STA access to the network. The communication protocol between the AP and the authentication server is not specified by 802.11i but is specified by WPA to be RADIUS. Most deployments of 802.11 would probably end up using RADIUS. The RADIUS protocol does allow for securely distributing the key from the authentication server to the AP and this is how the PMK gets to the AP.

Note that 802.1X is a framework for authentication. It does not specify the authentication protocol to be used. Therefore, it is up to the network administrator to choose the authentication protocol (s)he wants to "plug-in" to the 802.1X architecture. One of the most often discussed authentication protocols to be used with 802.1X is TLS. Figure 7.13 summarizes how TLS can be used as an authentication protocol in a EAP over WLAN environment. The EAP-TLS protocol is well documented. It has been analyzed extensively and no significant weaknesses have been found in the protocol itself. This makes it an attractive option for security use in 802.1X. However, there is a deployment issue with this scheme.

Note that EAP-TLS relies on certificates to authenticate the network to the clients and the clients to the networks. Requiring the network (i.e., the servers) to have certificates is a common theme in most security architectures. However, the requirement that each client be issued a certificate leads to the requirement of the widespread deployment of PKI. Since this is sometimes not a cost-effective option, a few alternative protocols have been proposed: EAP-TTLS (*Tunneled TLS*) and PEAP. Both of these protocols use certificates to authenticate the network (i.e., the server) to the client but do not use certificates to authenticate the client to the server. This means that a client no longer needs a certificate to authenticate itself to the server; instead, the client can use password-based schemes (CHAP, PAP, etc.) to authenticate

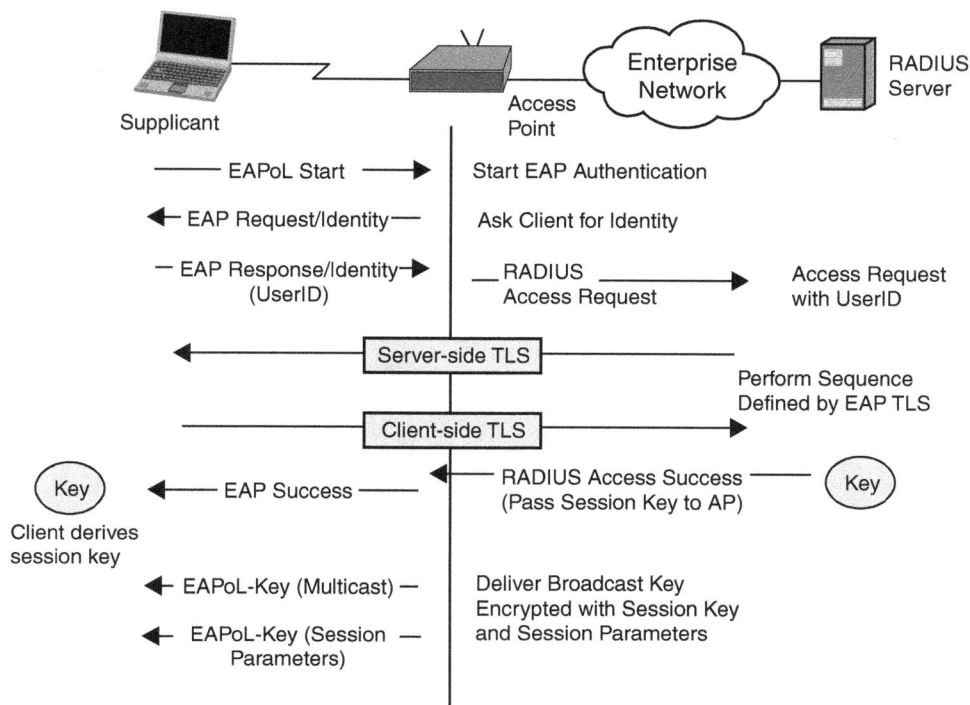

Figure 7.13: 802.1X

themselves. Both protocols divide the authentication process into two phases. In phase 1, authenticate the network (i.e., the server) to the client using a certificate and establish a *TLS tunnel* between the server and the client. This secure[21] TLS channel is then used to carry out a password-based authentication protocol to authenticate the client to the network (server).

7.8.3 Confidentiality

Recall from section 7.5.1 that the fundamental WEP loophole stems from using a stream cipher in an environment filled with packet loss. To work around this problem, WEP designers changed the encryption key for each packet. To generate the per-packet encryption key, the IV was concatenated with the preshared key. Since the preshared key is fixed, it is the IV which is used to make each per-packet key unique. There are multiple problems with this approach.

First, the IV size at 24 bits is too short. At 24 bits there are only 17,777,217 values before a duplicate IV value is used. Second, WEP does not specify how to select an IV for each

---

[21] Secure since it protects the identity of the client during the authentication process.

packet.[22] Third, WEP does not even make it mandatory to vary the IV on a per-packet basis—this means WEP explicitly allows reuse of per-packet keys. Fourth, there is no mechanism to ensure the IV is unique on a per-station basis. This makes the IV collision space shared between stations, thus making a collision even more likely. Finally, simply concatenating the IV with the preshared key to obtain a per-packet key is cryptographically insecure, making WEP vulnerable to the FMS attack. The FMS attack exploits the fact that the WEP creates the per-packet key by simply concatenating the IV with the master-key. Since the first 24 bits of each per-packet key is the IV (which is available in plain-text to an eavesdropper[23]), the probability of using "weak-keys"[24] is very high.

First off, TKIP doubles the IV size from 24 bits to 48 bits. This results in increasing the "time to key collision" from a few hours to a few hundred years. Actually, the IV is increased from 24 bits to 57 bits by requiring the insertion of 32 bits between the existing WEP IV and the start of the encrypted data in the WEP packet format. However, only 48 bits of the IV are used since 8 bits are reserved for discarding some known (and some yet to be discovered) weak-keys.

Simply increasing the IV length will, however, not work with the existing WEP hardware accelerators. Remember that existing WEP hardware accelerators expect a 24-bit IV as an input to concatenate with a preshared key (40/104 bit) in order to generate the per-packet key (74/128 bit). This hardware cannot be upgraded to deal with a 48-bit IV and generate an 88/157-bit key. The approach, therefore, is to use per-packet key mixing as explained in section 7.8.1. Using the per-packet key-mixing function (much more complicated) instead of simply concatenating the IV to the master key to generate the per-packet key increases the effective IV size (and hence improves on WEP security) while still being compatible with existing WEP hardware.

7.8.4 Integrity

WEP used CRC-32 as an integrity check. The problem with this protocol was that it was linear. As we saw in section 7.6, this is not a cryptographically secure integrity protocol. It does, however, have the merit that it is not computation intensive. What TKIP aims to do is to specify an integrity protocol that is cryptographically secure and yet not computation intensive, so it can be used on existing WEP hardware that has very little computation power. The

---

[22] Implementations vary from a sequential increase starting from zero to generating a random IV for each packet.
[23] Remember that each WEP packet carries the IV in plain-text format prepended to the encrypted packet.
[24] Use of certain key values leads to a situation where the first few bytes of the output are not all that random. Such keys are known as weak-keys. The simplest example is a key value of 0.

problem is that most well-known protocols used for calculating a MIC (message integrity check) have lots of multiplication operations and multiplication operations are computation intensive. Therefore, TKIP uses a new MIC protocol—*MICHAEL*—which uses no multiplication operations and relies instead on shift and add operations. Since these operations require much less computation, they can be implemented on existing 802.11 hardware equipment without affecting performance.

Note that the MIC value is added to the MPDU in addition to the ICV that results from the CRC32. It is also important to realize that MICHAEL is a compromise. It does well to improve upon the linear CRC-32 integrity protocol proposed in WEP while still operating within the constraints of the limited computation power. However, it is in no way as cryptographically secure as the other standardized MIC protocols such as MD5 or SHA-1. The TKIP designers knew this and hence built in countermeasures to handle cases where MICHAEL might be compromised. If a TKIP implementation detects two failed forgeries (i.e., two packets where the calculated MIC does not match the attached MIC) in one second, the STA assumes that it is under attack and as a countermeasure deletes its keys, disassociates, waits for a minute and then reassociates. Even though this may sound a little harsh since it disrupts communication, it does avoid forgery attacks.

Another enhancement that TKIP makes in IV selection and use is to use the IV as a sequence counter. Recall that WEP did not specify how to generate a per-packet IV.[25] TKIP explicitly requires that each STA start using an IV with a value of 0 and increment the value by one for each packet that it transmits during its session[26] lifetime. This is the reason the IV can also be used as a TSC (*TKIP sequence counter*). The advantage of using the IV as a TSC is to avoid the replay attack to which WEP was susceptible.

TKIP achieves replay protection by using a unique IV with each packet that it transmits during a session. This means that, in a session, each new packet coming from a certain MAC address would have a unique number.[27] If each packet from Alice had a unique number, Bob could tell when Eve was replaying old messages. WEP does not have replay protection since it cannot use the IV as a counter. Why? Because WEP does not specify how to change IV from one packet to another and, as we saw earlier, it does not even specify that you need to.

---

[25] In fact, WEP did not even specify that the IV had to be changed on a per-packet basis.
[26] An 802.11 session refers to the association between a STA and an AP.
[27] At least for 900 years—that's when the IV rolls over.

7.8.5 The Overall Picture: Confidentiality + Integrity

The overall picture of providing confidentiality and message integrity are shown in Figure 7.14.

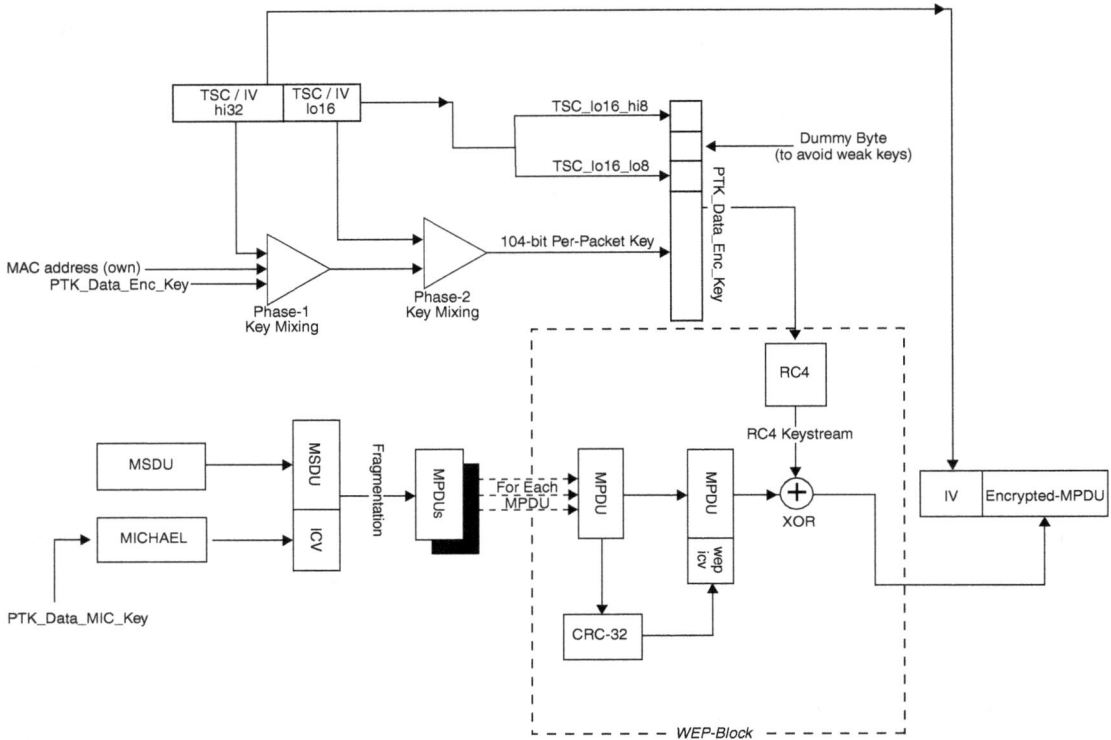

Figure 7.14: WPA Confidentiality and Integrity

7.8.6 How WPA Fixes WEP Loopholes

In section 7.7 we summarized the loopholes of WEP. At the beginning of section 7.8 we said that WPA/TKIP was designed to close these loopholes while still being able to work with existing WEP hardware. In the table below, we summarize what all WPA/TKIP achieves and how.

Table 7.1: WEP vs. WPA

WEP	WPA
Relies on preshared, aka (out-of-band) key establishment mechanisms. Usually leads to manual configuration of keys and to key sharing among STAs in a BSS (often ESS).	Recommends 802.1X for authentication and key establishment in enterprise deployments. Also supports preshared key establishment like WEP.
Uses a synchronous stream cipher which is unsuitable for the wireless medium.	Same as WEP.
Generates per-packet key by concatenating the IV directly to the master/pre-shared key, thus exposing the base-key/master-key to attacks like FMS.	Solves this problem by (a) introducing the concept of PTK in the key hierarchy and (b) by using a key-mixing function instead of simple concatenation to generate per-packet keys. This reduces the exposure of the master key.
Static master key + small size of IV + method of per-packet key generation → Extremely limited key space.	Increases the IV size to 57 bits and uses only 48 of these bits, reserving 8 bits to discard weak-keys. Also, use of PTK, which are generated afresh for each new session, increases the effective key space.
Changing the IV with each packet is optional → key reuse highly probable.	Explicitly specifies that both the transmitter and the receiver initialize the IV to zero whenever a new set of PTK is established[28] and then increment it by one for each packet it sends.
Linear algorithm (CRC-32) used for message integrity → Weak integrity protection.	Replaces the integrity-check algorithm to use MICHAEL, which is nonlinear. Also, specifies countermeasures for the case where MICHAEL may be violated.
ICV does not protect the integrity of the 802.11 header → Susceptible to redirection attacks.	Extends the ICV computation to include the MAC source and destination address to protect against redirection attacks.
No protection against replay attacks.	The use of IV as a sequence number provides replay protection.
No support for a STA to authenticate the network.	Use of 802.1X in enterprise deployments may allow for this.

7.9 WPA2 (802.11i)

Recall from section 7.8 that WPA (Wi-Fi Protected Access) was specified by the Wi-Fi Alliance with the primary aim of enhancing the security of existing 802.11 networks by designing a solution that could be deployed with a simple software (or firmware) upgrade and without the need for a hardware upgrade. In other words, WPA was a stepping-stone to the final

---

[28] This usually happens every time the STA associates with an AP.

solution that was being designed by the IEEE 802.11i task group. This security proposal was referred to as the RSN (robust security network) and also came to be known as the 802.11i security solution. The Wi-Fi Alliance integrated this solution in their proposal and called it WPA2. We look at this security proposal in this section.

7.9.1 Key Establishment

WPA was a prestandard subset of IEEE 802.11i. It adopted the key establishment, key hierarchy and authentication recommendations of 802.11i almost completely. Since WPA2 and 802.11i standard are the same, the key-establishment process and the key-hierarchy architecture in WPA and WPA2 are almost identical. There is one significant difference though. In WPA2, the same key can be used for the encryption and integrity protection of data. Therefore, there is one less key needed in WPA2. For a detailed explanation of how the key hierarchy is established, see section 7.8.1.

7.9.2 Authentication

Just like key establishment and key hierarchy, WPA had also adopted the authentication architecture specified in 802.11i completely. Therefore, the authentication architecture in WPA and WPA2 is identical. For a detailed explanation of the authentication architecture, see section 7.8.2.

7.9.3 Confidentiality

In this section we look at the confidentiality mechanism of WPA2 (802.11i). Recall that the encryption algorithm used in WEP was RC4, a stream cipher. Some of the primary weaknesses in WEP stemmed from using a stream cipher in an environment where it was difficult to provide lossless synchronous transmission. It was for this reason that Task Group-i specified the use of a block encryption algorithm when redesigning 802.11 security. Since AES was (and still is) considered the most secure block cipher, it was an obvious choice. This was a major security enhancement since the encryption algorithm lies at the heart of providing confidentiality.

Realize that specifying an encryption algorithm is not enough for providing system security. What is also needed is to specify a "mode" of operation. To provide confidentiality in 802.11i, AES is used in the counter mode. Counter mode actually uses a block cipher as a stream cipher, thus combining the security of a block cipher with the ease of use of a stream cipher. Figure 7.15 shows how AES counter mode works.

Using the counter mode requires a counter. The counter starts at an arbitrary but predetermined value and is incremented in a specified fashion. The simplest counter operation, for example, starts the counter with an initial value of 1 and increments it sequentially by 1 for each block. Most implementations, however, derive the initial value of the counter from

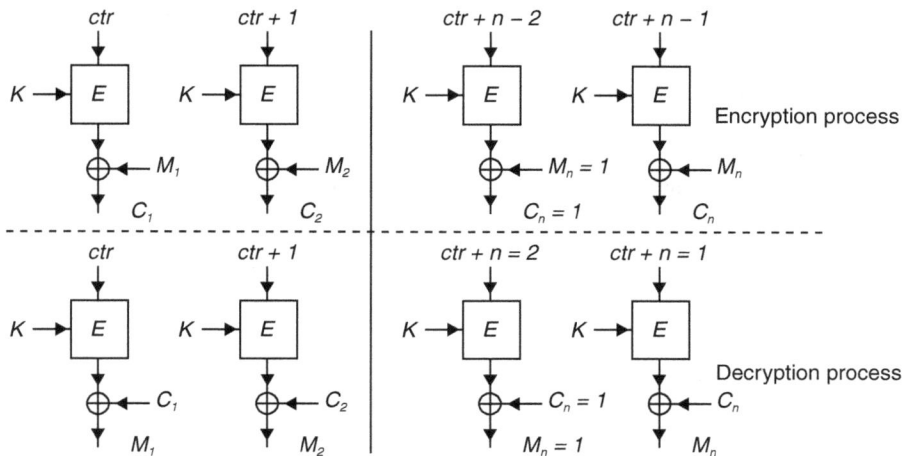

Figure 7.15: AES Counter Mode

a Nonce value that changes for each successive message. The AES cipher is then used to encrypt the counter to produce a "key stream." When the original message arrives, it is broken up into 128-bit blocks and each block is XORed with the corresponding 128 bits of the generated key stream to produce the cipher text.

Mathematically, the encryption process can be represented as $C_i = M_i (+) E_k(i)$ where i is the counter. The security of the system lies in the counter. As long as the counter value is never repeated with the same key, the system is secure. In WPA2, this is achieved by using a fresh key for every session (see section, 7.8.1).

To summarize, the salient features of AES in counter mode are as follows:

1. It allows a block cipher to be operated as a stream cipher.

2. The use of counter mode makes the generated key stream independent of the message, thus allowing the key stream to be generated before the message arrives.

3. Since the protocol by itself does not create any interdependency between the encryption of the various blocks in a message, the various blocks of the message can be encrypted in parallel if the hardware has a bank of AES encryption engines.

4. Since the decryption process is exactly the same as encryption,[29] each device only needs to implement the AES encryption block.

---

[29] XORing the same value twice leads back to the original value.

5. Since the counter mode does not require that the message be broken up into an exact number of blocks, the length of the encrypted text can be exactly the same as the length of the plain-text message.

Note that the AES counter mode provides only for the confidentiality of the message and not the message integrity. We see how AES is used for providing the message integrity in the next section. Also, since the encryption and integrity protection processes are very closely tied together in WPA2/802.11i, we look at the overall picture after we have discussed the integrity process.

7.9.4 Integrity

To achieve message integrity, Task Group-i extended the counter mode to include a CBC MAC operation. This is what explains the name of the protocol: AES-CCMP where CCMP stand for *counter mode CBC MAC* protocol. The CBC-MAC protocol (also known as CBC-Residue) is reproduced here in Figure 7.16 where the black boxes represent the encryption protocol (AES in our case).

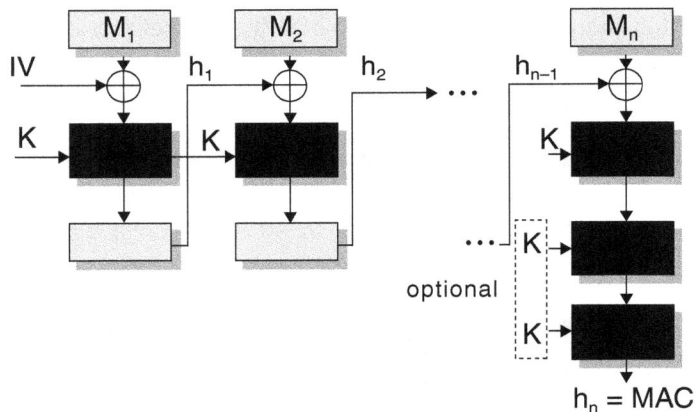

Figure 7.16: CBC MAC

As shown in the figure, CBC-MAC keeps a running XOR of all the plain-text blocks and XORs that with the last plain-text block before encrypting it. This ensures that any change made to any cipher text (for example, by a malicious intruder) block changes the decrypted output of the last block. CBC-MAC is an established technique for message integrity. What Task Group-i did was to combine the counter mode of operation with the CBC-MAC integrity protocol to create the CCMP. A nice side benefit of using AES as the core in both the encryption and integrity portions of the algorithm is that 802.11 chip sets can re-use the same AES hardware accelerator engine for both, thus saving silicon real estate.

7.9.5 The Overall Picture: Confidentiality + Integrity

Since a single process is used to achieve integrity and confidentiality, the same key can be used for the encryption and integrity protection of data. It is for this reason that there is one less key needed in WPA2. The complete process, which combines the counter-mode encryption and CBC-MAC integrity, works as follows.

In WPA2, the PTK is 384 bits long. Of this, the most significant 257 bits form the EAPoL MIC key and EAPOL encryption key. The least significant 128 bits form the data key. This data key is used for both encryption and integrity protection of the data. Before the integrity protection or the encryption process starts, a CCMP header is added to the 802.11 packet before transmission. The CCMP header is 8 bytes in size. Of these 8 bytes, 7 bytes are used for carrying the *packet number* (PN), which is needed for the other (remote) end to decrypt the packet and to verify the integrity of the packet. One byte is reserved for future use and the remaining byte contains the key-ID. Note that the CCMP header is prepended to the payload of the packet and is not encrypted since the remote end needs to know the PN before it starts the decryption or the verification process. The PN is a per-packet sequence number that is incremented for each packet processed.

The integrity protection starts with the generation of an IV (*initialization vector*) for the CBC MAC process. This IV is created by the concatenation of the following entities: Flag, Priority, Source MAC address, a packet number (PN) and DLen as shown in Figure 7.17.

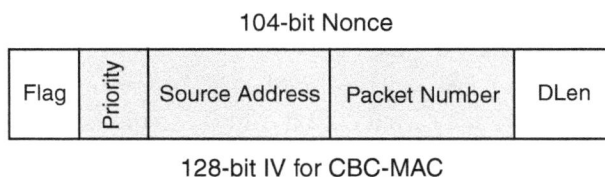

104-bit Nonce

Flag	Priority	Source Address	Packet Number	DLen

128-bit IV for CBC-MAC

Figure 7.17: Initialization Vector in WPA2

The *flag* field has a fixed value of 01011001. The *priority* field is reserved for future use. The source MAC address is self explanatory and the packet number (PN) is as we discussed above. Finally, the last entity *DLen* indicates the *data length* of the plain text. Note that the total length of the IV is 128 bits and the priority, source address and the packet number fields together also form the 104-bit *Nonce* (shaded portion of Figure 7.17), which is required in the encryption process. The 128-bit IV forms the first block, which is needed to start the CBC-MAC process described in section 7.8.3. The CBC-MAC computation is done over the 802.11 header and the MPDU payload. This means that this integrity-protection scheme also protects the source and the destination MAC address, the QoS traffic class and the data length. Integrity protecting the header along with the MPDU payload protects against replay attacks. Note that the CBC-MAC process requires an exact number of blocks to operate on. If the length of the plain-text data cannot be divided into an exact number of blocks, the plain-text data needs to be added for the purposes of MIC computation.

Once the MAC has been calculated and appended to the MPDU, it is now ready for encryption. It is important to re-emphasize that only the data part and the MAC part of the packet are encrypted, whereas the 802.11 header and the CCMP header are not encrypted. From section 7.8.2, we know that the AES-counter mode encryption process requires a key and a counter. The key is derived from the PTK as we discussed. The counter is created by the concatenation of the following entities: Flag, Priority, Source MAC address, a packet number (PN) and Ctr as shown in Figure 7.18.

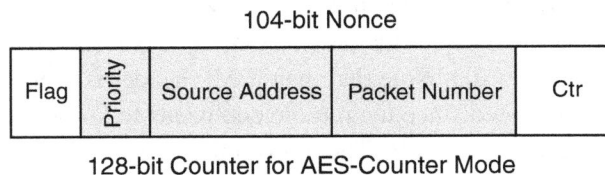

104-bit Nonce

Flag	Priority	Source Address	Packet Number	Ctr

128-bit Counter for AES-Counter Mode

Figure 7.18: Counter in WPA2

Comparing Figure 7.18 with Figure 7.17, we see that the IV for the integrity process and the counter for the encryption process are identical except for the last 17 bits. Whereas the IV has the last 17 bits as the length of the plain text, the counter has the last 17 bits as Ctr. It is this Ctr that makes the counter a real "counter." The value of Ctr starts at 1 and counts up as the counter mode proceeds. Since the Ctr value is 17 bits, this allows for up to 2^{17} (75,537) blocks of data in a MPDU. Given that AES uses 128-bit blocks, this means that an MPDU can be as long as 2^{23}, which is much more than what 802.11 allows: so the encryption process does not impose any additional restrictions on the length of the MPDU.

Even though CCMP succeeds in combining the encryption and integrity protocol in one "process," it does so at some cost. First, the encryption of the various message blocks can no longer be carried out in parallel since CBC-MAC requires the output of the previous block to calculate the MAC for the current block. This "slows down" the protocol. Second, CBC-MAC requires the message to be broken into an exact number of blocks. This means that if the message cannot be broken into an exact number of blocks, we need to add padding bytes to it to do so. The padding technique has raised some security concerns among some cryptographers but no concrete threats have been identified at present.

The details of the overall CCMP are shown in Figure 7.19 and finally Table 7.2 compares the WEP, WPA and WPA2 security architectures.

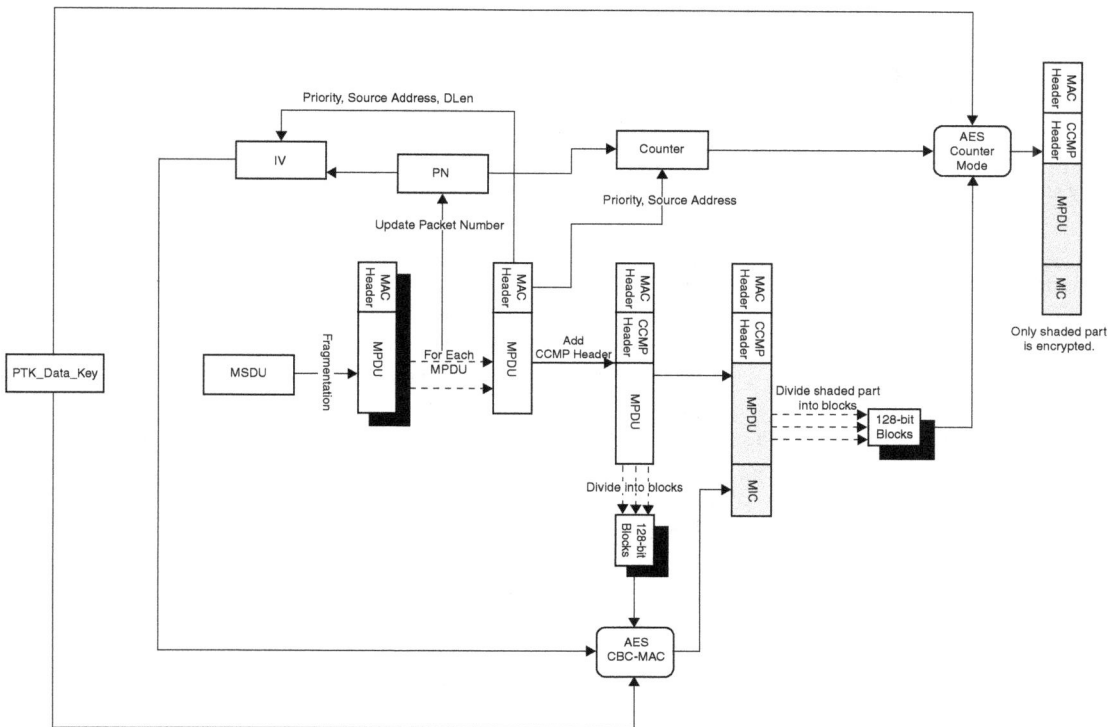

Figure 7.19: WPA2 Architecture

Table 7.2: Wep, WPA, WPA2 Comparison

WEP	WPA	WPA2
Relies on preshared aka (out-of-band) key-establishment mechanisms. Usually leads to manual configuration of keys and to key sharing among STAs in a BSS (often ESS).	Recommends 802.1X for authentication and key establishment in enterprise deployments. Also supports preshared key establishment like WEP.	Same as WPA.
Uses a synchronous stream cipher which is unsuitable for the wireless medium.	Same as WEP.	Replace a stream cipher (RC4) with a strong block cipher (AES).
Generates per-packet key by concatenating the IV directly to the master/preshared key, thus exposing the base-key/master-key to attacks like FMS.	Solves this problem (a) by introducing the concept of PTK in the key hierarchy and (b) by using a key mixing function instead of simple concatenation to generate per-packet keys. This reduces the exposure of the master key.	Same as WPA.

Table 7.2: Wep, WPA, WPA2 Comparison (continued)

WEP	WPA	WPA2
Static master key + Small size of IV + Method of per-packet key generation → Extremely limited key space.	Increases the IV size to 57 bits and uses only 48 of these bits reserving 8 bits to discard weak-keys. Also, use of PTK, which are generated afresh for each new session increases the effective key space.	Same as WPA.
Changing the IV with each packet is optional → key reuse highly probable.	Explicitly specifies that both the transmitter and the receiver initialize the IV to zero whenever a new set of PTK is established[30] and then increment it by one for each packet it sends.	Same as WPA.
Linear algorithm (CRC-32) used for message integrity → Weak integrity protection.	Replaces the integrity check algorithm to use MICHAEL which is nonlinear. Also, specifies countermeasures for the case where MICHAEL may be violated.	Provides for stronger integrity protection using AES-based CCMP.
ICV does not protect the integrity of the 802.11 header → Susceptible to Redirection Attacks.	Extends the ICV computation to include the MAC source and destination address to protect against redirection attacks.	Same as WPA.
No protection against replay attacks.	The use of IV as a sequence number provides replay protection.	Same as WPA.
No support for a STA to authenticate the network.	No explicit attempt to solve this problem but the recommended use of 802.1X could be used by the STA to authenticate the network.	Same as WPA.

7.10 Beyond 802.11 Security

Till now, we have discussed the security aspects provided for by the 802.11 standard. However, since 802.11 is a Layer-2 standard, these measures operate at the link level (of the seven-layer ISO model) and are responsible only for securing the link between a wireless STA and the AP. In other words, they do not provide end-to-end security, which is needed for voice (both signaling and media stream). Thus, additional mechanisms are necessary for truly secure voice communication. In this section we will discuss some of the options for these additional security layers.

---

[30] This usually happens every time the STA associates with an AP.

7.10.1 IPsec: Security at Layer 3

Unlike Layer 2 which provides link-layer connectivity, Layer 3 is responsible for providing end-to-end connectivity. IPsec is a mechanism to provide end-to-end security in IP networks. It operates effectively at the bottom of the IP network layer, and is implemented as a component at the bottom of the IP protocol stack (as illustrated in Figure 7.20).

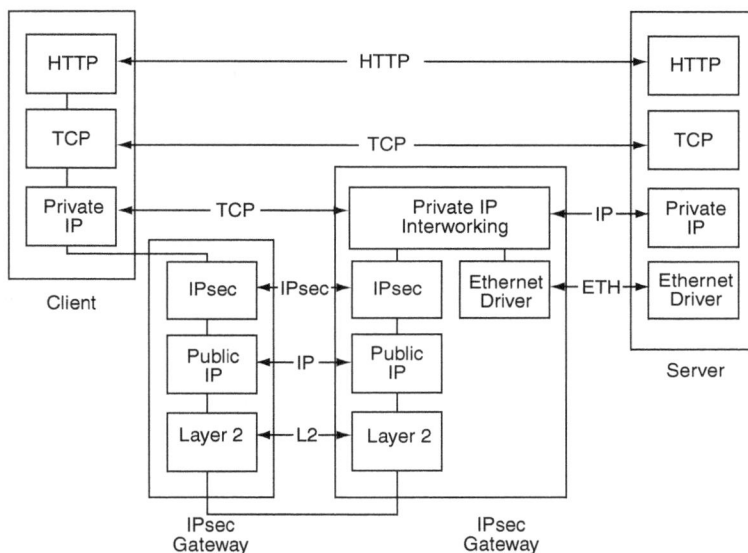

Figure 7.20: IPsec Architecture

IPsec is a suite of protocols, specified in several RFCs. RFC 2406 defines the main Encapsulating Security Protocol (ESP). RFC 2402 defines an adjunct authentication protocol referred to as Authentication Header (AH). Additional RFCs (e.g., RFC 3602, 3566, 3686, 2403, 2405) define profiles for how various cryptographic techniques are to be used. IPsec is designed to work primarily with a key-management protocol known as the Internet Key Exchange (IKE) defined in its own RFC 2409. However, IPsec can be used with other key-management protocols (such as Kerberos [RFC 1510]).

7.10.1.1 Security Associations in IPsec

Remember that the aim of IPsec is to protect the communication between two IP addresses. However, one IP address might be communicating with more than one IP address at any given time and each one of these sessions might be using IPsec. Furthermore, each of these IPsec sessions may be using separate security parameters. How, then, do we distinguish between the different IPsec sessions, given that all of them are using the same IP address? To achieve this, each system implementing IPsec maintains a security association database.

The term *security association* (SA) refers to a cryptographically protected connection.

An SA is unidirectional in that it specifies completely all the cryptographic information required in one direction of the communication. This includes information like the identity of the remote end, the cryptographic key being used, the cryptographic services and algorithms in use, the sequence number currently in use, and so on. Each entry in the SA database contains the SAs that are currently being used by this system. These entries are indexed by the destination address and the security parameter index (SPI). The system searches the database using the destination address when it needs to transmit a packet to that destination address using IPsec. On the other hand, when the system receives an IP packet, it examines the AH/ESP header to determine the SPI and then uses this SPI to index into the SA database to find out all the information it needs to process this packet.

7.10.1.2 IPsec Packet Format

AH provides support only for integrity protection of the payload in the packet and for some fields in the IP header itself. On the other hand, ESP provides for encryption and/or integrity protection only for the payload in the packet. As shown in Figure 7.21, most of the fields like the SPI, sequence number and payload length are common between the AH and ESP header. The most prominent difference is padding. ESP requires padding whereas AH does not. This is because AH provides support only for integrity protection whereas ESP provides support for encryption along with integrity protection. Since most encryption ciphers require the data to be an integral number of bytes in length, padding is supported in the ESP header. The zero or more padding bytes are used to get the packet aligned to a 32-bit boundary.

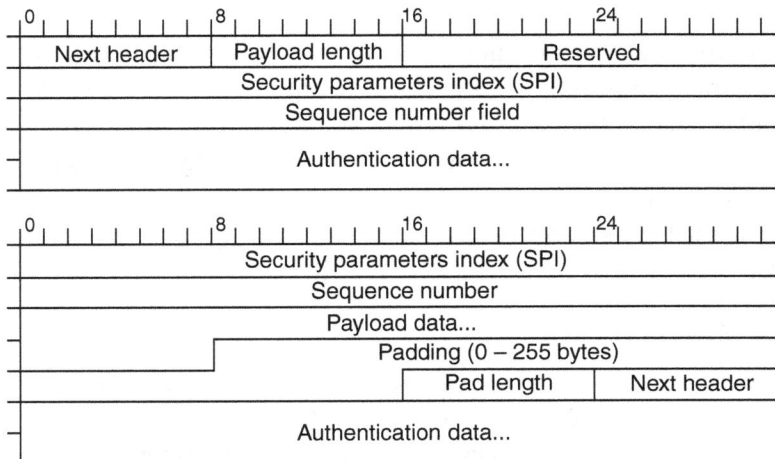

Figure 7.21: Authentication Header and ESP Packet Formats

The ESP support for encryption also leads to another difference, which at first is not so obvious from the header format. Note that IPsec is not bound to any encryption or authentication algorithm and can support several algorithms for encryption (AES, 3DES, IDEA, Blowfish) and authentication (MD5, SHA1). Now, if the encryption algorithm used to encrypt the payload in ESP requires cryptographic synchronization data—for example, an initialization vector (IV)—then this data MAY be carried explicitly in the Payload field of the ESP header. For example, if the encryption algorithm used is "AES in CBC mode," then an IV of 16 bytes is carried as part of the Payload fields. This IV is then used to seed CBC mode operation. Note that, unlike WPA or WPA2, the entire IV must be sent with every packet. IPsec does not make use of existing fields in the IP header to construct a per-packet IV.

7.10.1.3 IPsec Modes

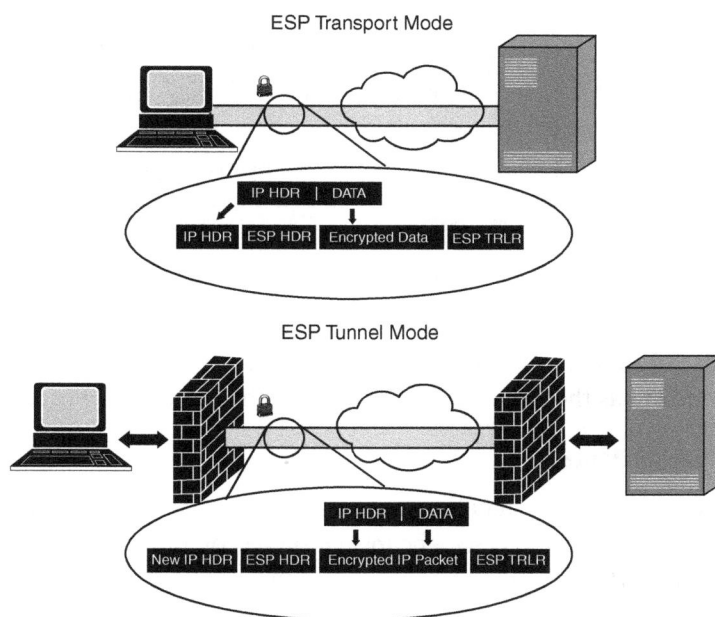

ESP Transport Mode

| IP HDR | DATA |

| IP HDR | ESP HDR | Encrypted Data | ESP TRLR |

ESP Tunnel Mode

| IP HDR | DATA |

| New IP HDR | ESP HDR | Encrypted IP Packet | ESP TRLR |

Figure 7.22: IPsec Modes

Now that we know what IPsec headers look like, the next question is, who inserts them? The obvious answer is—the system implementing IPsec. However, this requires a little more explanation. IPsec may be used by communication endpoints but also by intranet gateways to provide a secure "tunnel" between two intranets (secure environment) over the Internet (unsecure environment). In the former case, the IPsec headers are inserted between the IP header and the packet payload/data. In the tunneling case, however, the IPsec header is inserted around the IP packet and another new IP header is inserted around this encapsulated packet. The new IP header contains the IP addresses of the endpoints of the tunnel, since it is at the remote gateway that this external IP header and the IPsec header are removed to restore the old packet. Even though the tunneling encapsulation is costly in terms of overhead incurred, it is a very attractive proposition for corporate solutions that wish to connect intranets (think VPNs) without having to install IPsec at each and every client in the network.

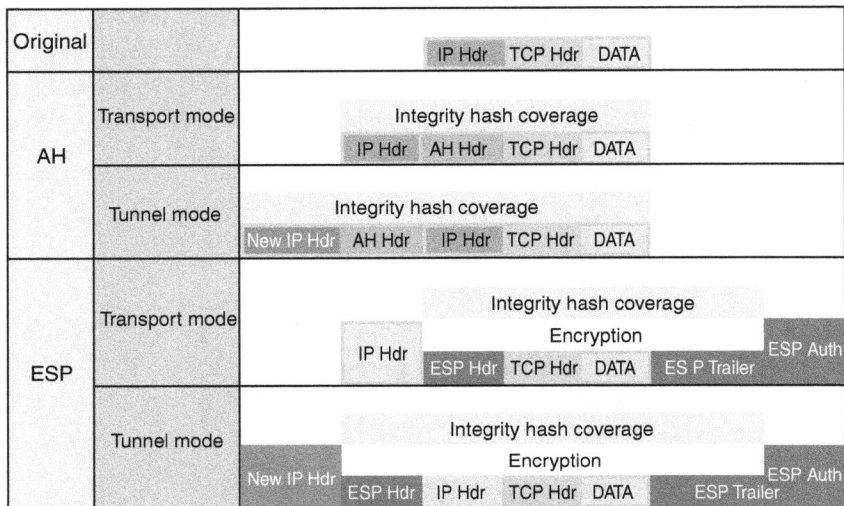

Figure 7.23: IPsec Packet Formats

Note that in the case of tunnel mode, the *original* IP header is encapsulated in a new IP header. So tunnel mode has the advantage of protecting the original IP header as well.

7.10.1.4 Using IPsec for Voice

IPsec can be used to protect both signaling and RTP media. However, for media it is important to note what the impact of the IPsec encapsulation has on the overall media packet size (especially due to the ESP padding requirements and the requirement to send the initialization vector with each packet). This is illustrated in Table 7.3. Briefly, the use of IPsec diminishes the benefit of low-bit-rate codecs as a means of reducing bandwidth.

Table 7.3: Impact of IPsec ESP Mode on Voice Packet Size

Packet Breakdown (all sizes in bytes)	G729a 20 msc (20 bytes per packet)	G711, 20 msc (160 bytes per packet)
Payload	20	160
RTP Header	12	12
UDP	8	8
IP	20	20
Baseline Voice Packet Size	60	220
2nd IP Header (Tunnel Mode)	20	20
ESP Header	4	4
Initialization Vector (3DES/AES)	8 / 16	8 / 16
Padding (3DES/AES)	2/2	0/0
ESD Trailer	2	2
Authentication	12	12
IPsec Voice Packet Size (3DES/AES)	108/116	266 / 274
% Increase (3DES/AES)	**80 % / 93 %**	**20 % / 25 %**

7.10.2 TLS: Security at Layer 4

Secure Sockets Layer (SSL) and Transport Layer Security (TLS) are slightly incompatible protocols but are similar enough to be described together. The SSL/TLS protocol is based on the philosophy that it is easier to modify the application than the operating system. Therefore, SSL sits between the application layer and the transport layer, and applications interface to the SSL layer rather than the transport layer. From a programming viewpoint, applications use SSL sockets rather than transport layer sockets for network programming.

The name of the protocol is therefore self-explanatory. SSL is a layer (a library of wrapper functions) around the socket interface. It runs as a user process on top of transmission control protocol (TCP) (port number 443). Applications that wish to use SSL/TLS should use the SSL library functions for networking instead of directly calling the socket functions. This allows the two communicating parties to authenticate and establish a session key that is used to protect the remainder of the session cryptographically. The obvious assumption here is that both the communicating applications must be using SSL/TLS.

Figure 7.24 shows how the client, Alice, authenticates with the server Bob and establishes a secret (referred to as the premaster secret). The TLS handshake begins with Alice sending a Client-Hello message to Bob, though this may in some cases be preceded by the server (Bob) prompting the client (Alice) to send a Client-Hello message using the Hello-Request message. The Client-Hello message sent by Alice contains, among other things, a random number generated by Alice, the Session-Id and a list of cipher-suites supported by Alice. The Session-Id is Null for a new connection but may be used by Alice for session resumption in the near future.

Alice   Client Hello (R$_A$, SID, SuCS)   Bob

Server Hello (R$_B$, SID, SeCS)

Server Cert. (Certificate :: g, n, g^x mod n)

Server Hello Done ()

Client Key Exchange (g^x mod n)

Pre-master secret established   Pre-master secret established

Change CipherSpec

Finished (PRF(master_secret, "client finished", MD5(handshake_msgs) + SHA1(handshake_msgs)))

Change CipherSpec

Finished (PRF(master_secret, "server finished", MD5(handshake_msgs) + SHA1(handshake_msgs)))

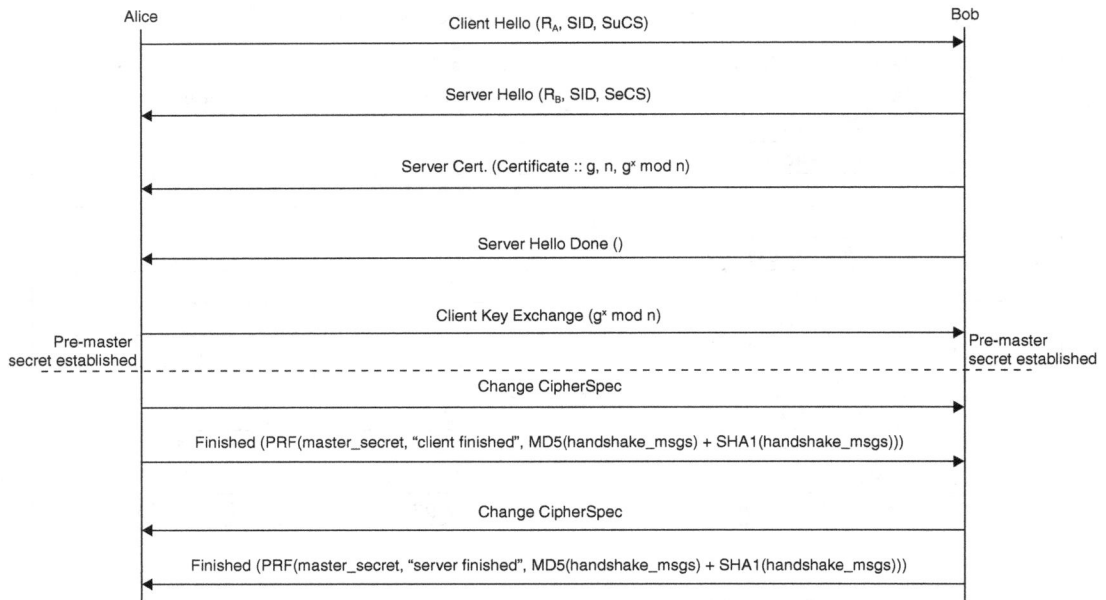

Figure 7.24: TLS Protocol Handshake

On receiving the Client-Hello message from Alice, Bob responds with a Server-Hello message which contains, among other things, a random number generated by Bob, the Session-Id and a selection from the list of cipher-suites that Alice supplied in the Client-Hello message. If the Session-Id in the Client-Hello message was Null—that is, if this is a new connection—Bob generates a Session-Id and returns it in this message. If, on the other hand, this is a session-resumption attempt from Alice, the Session-Id would be non-Null. In this case, Bob would look up its session cache to find this Session-Id. If a match is found and Bob is willing to establish the new connection using the previously established parameters, Bob would send back a Finished message with the same Session-Id that it received in the Client-Hello message. If Bob does not find a matching Session-Id in its cache or if Bob is not willing to resume the session, Bob will send back a Server-Hello message with a different Session-Id.

The Server-Hello message from Bob is followed by a Server-Certificate message that contains a certificate type appropriate to the selected cipher-suites' key-exchange algorithm. Figure 7.24 assumes that the DH key-exchange algorithm is in use. In this case, the Server-Certificate is followed by a Server-Hello-Done message. Note that some key-exchange algorithms (like RSA) may require an additional Server-Key-Exchange message to be sent before the Server-Hello-Done message.

On receiving the Server-Hello-Done message from Bob, Alice sends the Client-Key-Exchange message to Bob. In the DH case, this message contains Alice's public key. After this message, both Alice and Bob can derive the premaster secret independently without worrying about Eve possessing this secret. For why this is so, see sections 2.2.1 and 2.2.2. The premaster secret is then converted to a master secret using the pseudo-random-function, the client random number and the server random number; that is:

master_secret = PRF(pre_master_secret, "master secret", ClientHello.random+ServerHello.random)

where "+" denotes a concatenation. The pseudo-random-function for TLS version 1.0 is created by splitting the premaster secret into two halves and using one half to generate data with the MD5 MAC and the other half to generate data with the SHA1 MAC. The Client-Key-Exchange message from Alice is followed by the Change-Cipher-Spec message, which indicates to Bob that from now on Alice is going to use the derived master secret to protect the rest of the session. The first message protected with the just-negotiated algorithms, keys and secrets that Alice sends to Bob is the Finished message. The Finished message contains as data the signed hash of all the handshake messages exchanged between Alice and Bob up until now. This means that the recipient of the Finished message can use the data contained in this message to verify that a) the other end has derived the master secret correctly, and b) none of the handshake messages have been modified or spoofed by a malicious Eve. On receiving the Finished message from Alice, Bob also sends a Change-Cipher-Spec message followed by a Finished message. From now on, applications at both ends can start sending data securely.

To summarize, TLS uses the Hello messages to agree on cipher-suites (algorithms), exchange random numbers and check for session resumption. Then, the client and the server exchange certificates and cryptographic parameters to authenticate themselves and "independently" derive a premaster secret. Using this premaster secret and the random values exchanged earlier, both endpoints derive a master secret, which is used as a basis for protecting the rest of the session.

7.10.2.1 TLS and Voice

Just as HTTP traffic can be protected as "https" with Secure Sockets Layer (SSL), SIP signaling messages can be sent using "sips" transport. Note that, when using SSL at this layer along with 802.1X-TLS, we end up using TLS at effectively two levels. The use of TLS at effectively two levels in a Wi-Fi phone has several ramifications. The first is that the phone will need to support two security hierarchies or certificate trust chains. This is because the 802.11 network-security infrastructure will most likely be independent of the voice-provider infrastructure. The second is that, depending on how and where 802.11 security and voice security processing is distributed in the Wi-Fi phone, multiple copies of the same TLS software, TLS cipher suites, and security accelerators may be necessary. In the case of hardware accelerator modules, for example, the application of TLS to authenticate the phone and to protect key distribution has more stringent real-time requirements than protecting SIP signaling. Roaming times need to be on the order of tens of milliseconds, while SIP signaling can take hundreds of milliseconds.

7.10.3 SRTP

Unlike IPsec and TLS, the SRTP protocol, specified in RFC 3711, is a security protocol optimized for RTP media such as voice. The standard describes per-packet cryptographic techniques to encrypt and authenticate RTP media and the companion RTCP management stream. It does not specify any key management procedure, however. We will discuss various management techniques later in this section..

The format of a secure RTP packet is illustrated in Figure 7.25. SRTP adds two fields at the end of a normal RTP packet: the Master Key Index (MKI) and the authentication tag. These two do not increase the size of the RTP packet substantially. In particular, the authentication tag can be restricted to as little as four bytes. The MKI size is also negotiated and is optional. It can be used with a quasistatic key-management scheme where some number of master keys and "salts" (additional random strings used in some security algorithms) are downloaded to devices.

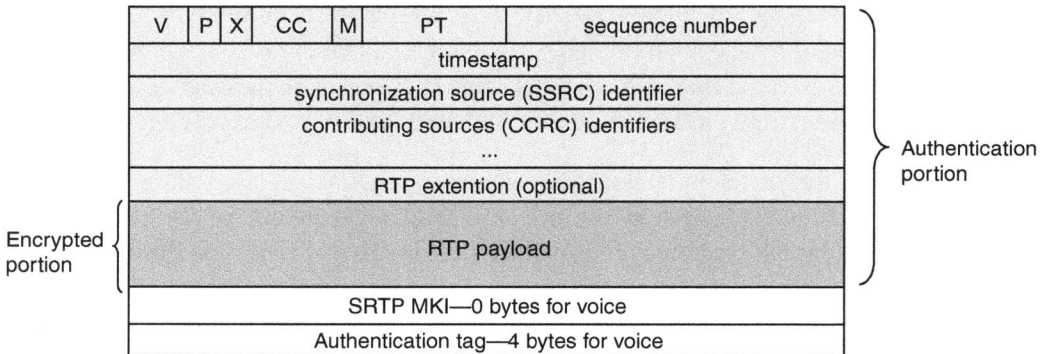

Figure 7.25: Secure RTP Packet Format

SRTP is open in the sense that new cryptographic algorithms can be added in the future. RFC 3711 defines two encryption algorithms. The first is AES block encryption in counter mode (the same mode used in WPA2). The second is AES encryption in F8 mode (an encryption mode developed in 3G UMTS networks). For authentication, SRTP mandates the HMAC-SHA algorithm (the same algorithm defined for IPsec). The number of authentication tag bytes to include can be negotiated at call setup.

Note that, unlike IPsec (which as a network-layer protocol cannot rely on knowledge of the application-level payload), SRTP does not require an initialization vector to be transported along with the packet. Instead, SRTP makes use of fields from the RTP header to construct a unique, per-packet IV as follows:

$$\text{IV} = (k\_s * 2^{16}) \text{ XOR } (\text{SSRC} * 2^{64}) \text{ XOR } (i * 2^{16}),$$

where:

    k_s = session salt key (derived from the master key and salt—see below)

    SSRC = ssrc in the RTP packet

    i = extended RTP sequence number.

This formulation requires both the sender and receiver to track an extended RTP sequence number. This is defined as the current sequence number plus the number of times the 16-bit RTP sequence number has rolled over, multiplied by 2^{16}. This gives us 2^{42} unique IVs; this is the maximum number of packets that can be protected with one session key.

SRTP also specifies the IV formation and algorithms to encrypt and authenticate RTCP packets. The keys and salts used in the algorithms are derived from a master key and salt as shown in Figure 7.26. The master key and salt come from external key-management procedures not defined by SRTP. These are fed into a pseudo-random function (PRF) that computes three outputs: the session encryption key, the session authentication key and the session salt. In the case of AES-128 counter mode, the session encryption and salt keys will be 128 bits long. The PRF for SRTP is also based on AES in counter mode. The current packet index can optionally feed into the key derivation. SRTP allows a key derivation rate to be specified (along with the master key and salt) through the key-management procedure. A setting of, say, 256000, will mean that the session keys and salt will be updated every 256K packets.

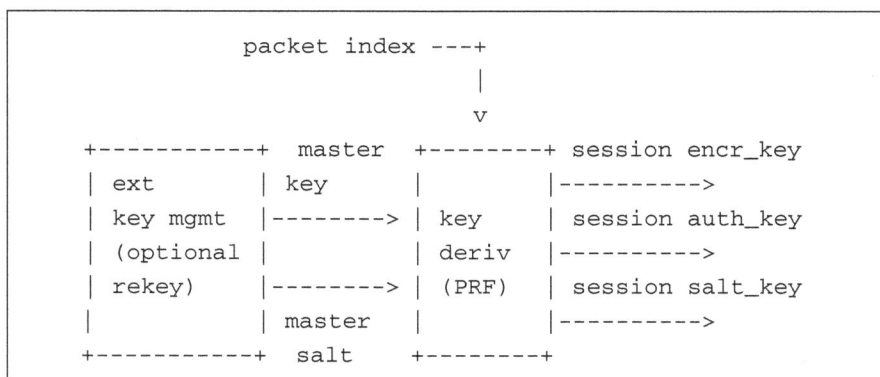

```
               packet index ---+
                               |
                               v
+-----------+  master  +--------+ session encr_key
| ext       | key      |        | ---------->
| key mgmt  |--------> | key    | session auth_key
| (optional |          | deriv  | ---------->
| rekey)    |--------> | (PRF)  | session salt_key
|           | master   |        | ---------->
+-----------+  salt    +--------+
```

Figure 7.26: Key Generation with SRTP

As mentioned above, SRTP key management is not yet standardized. There are several options being proposed.

- Preshared master keys and salts. Device provisioning, for example, could download one or more master key/salt pairs. The master key index feature of SRTP is used in the case of multiple provisioned master key/salts to switch masters on the fly.

- SDP-based, unprotected key exchange. There are several approaches. One proposal is illustrated in Figure 7.27. With this approach, the master key/salt and other SRTP session parameters are encoded (using base64) as SDP attribute lines (a=) and passed as part of call-signaling (i.e., SIP Invites and responses). This approach relies on the underlying call-signaling security layer (e.g., IPsec, TLS) to prevent the key/salt from being made public.

- MIKEY (RFC 3830). This draft standard embeds public key cryptography and certificates inside the call-signaling SDP. The MIKEY protocol is structured so that the public key handshake can take place within the normal SIP message sequence. However, it does require that each endpoint in the conversation maintain a public key certificate hierarchy and/or implement an efficient Diffie-Helman algorithm.

```
INVITE sip:user2@the.sipnet.com SIP/2.0 ..
   Max-Forwards: 70
   Route: <sip:proxy.the.sipnet.com;lr>
   ...
   v=0
   ..
   c=IN IP4 158.218.1.150
   .
   a=crypto: 1 AES_CM_128_HMAC_SHA1_80 inline: <key blob>
KDR=12
```

Figure 7.27: SDP-based SRTP Key Exhange

SRTP is a more efficient security protocol for media. However, its usage in conjunction with 802.11 has some of the same issues as discussed above for TLS. For example, the real-time requirements for an AES hardware accelerator module differ for the two cases. WPA2 link level operations require that the AES computations be done at near transmission speeds (e.g., up to 54 mbps).

7.11 Conclusion

802.11 security has been an area of widespread concern and discussion. The 802.11i protocol fixes many of the loopholes that were discovered in the base 802.11 security protocol. It is prudent for deployments to move to this new standard to ensure security of their network. For Voice-over-Wi-Fi, however, it is also important to remember that 802.11i security measures operate at Layer-2 standards and can therefore secure only the link between a wireless STA and the AP. Thus, additional mechanisms are necessary for truly secure voice communication. As described in this chapter, most applications will probably end up using IPsec or TLS/SSL for ensuring secure signaling and will use SRTP for a secure RTP stream in voice communication.

Roaming

8.1 The Need for Roaming

As section 5.8 explains, the basic need for roaming stems from attenuation of signals in the wireless domain. This attenuation explains why all wireless transmissions have a limited geographical range. This means that a mobile Wi-Fi phone, in most environments, will need to transition from one AP to another as it moves away from its current AP's range. We say "in most environments" since one application of VoWLAN is as a cordless phone replacement system with the AP acting as the base. It is possible that in such deployments the VoWLAN handset would be connected exclusively to its base AP without ever supporting or needing roaming. In an enterprise-like environment, however, there is usually a need to provide Wi-Fi coverage over a geographical area greater than the range provided by a single AP. Hence, roaming/handoffs[1] become necessary. In practice, deployments may further limit the range (aka cell size) of each access point to increase overall coverage and to prevent attached stations from adapting to a low transmission rate at the cell edge (and thus bring down the entire cell performance). Refer to the proposed transmit power control (TPC) of 802.11k draft for more on this. In summary, we can define roaming as the ability to find and associate with an AP when becoming (or about to become) disconnected from the current AP.

The basic idea behind roaming is illustrated in Figure 8.1. In the simplistic scenario, if we let the cell radius be R (ft), the user speed be S (ft/sec), and the degree of cell overlap be D (%), then we get the following formula for cell roam intervals, H (sec):

$$H = (2 * R * (1 - 2 * D/100)) / S$$

For example, with $R = 50$ ft, $S = 3$ miles/hr = 4.4 ft/sec, $D = 15\%$ we get $H = 19$ sec. That is, every 19 seconds the phone will need to roam to another AP. In this example, for a 2-minute call, with this simple model, a user would need around 6 AP-to-AP handoffs.

---

[1] The terms *roaming* and *handoffs* are used interchangeably throughout the text since in WLAN both refer to the act of disconnecting from one AP and connecting to another.

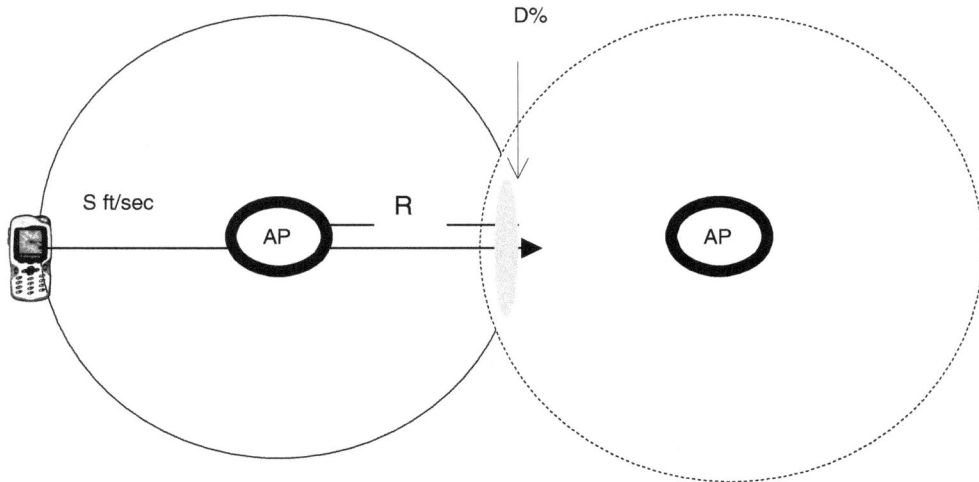

Figure 8.1: Basic Roaming Example (Intra-ESS)

However, this is a very simplistic model; a more realistic model of roaming in a wireless network would need to consider the following factors:

- Nonuniform construction in the building

- Nonuniform overlap between cells

- Variable motion

8.2 Types of Roaming

As defined, roaming/handoff in 802.11 refers to a STA disconnecting from one AP and connecting to another AP. However, within this definition, roaming may be classified into the following four types:

a. *Intra-ESS:* Handoffs between access points in the same ESS (as illustrated in Figure 8.1). In this case, the APs will share the same SSID and security implementation, and typically will handle networks with the same IP subnet (so that the phone's IP address does not need to change). This is the type of roaming that we will concentrate on in this chapter.

b. *Intra-ESS-with-SubnetChange:* Handoffs between APs in the same ESS (i.e., SSID enterprise network) but moving to a separate subnet. In this type of roaming, for example between buildings belonging to the same company, the APs will be responsible for handling the subnet mismatch, typically by using Virtual Private Network (VPN) tunneling to get the phone's traffic back to the home subnet. For our purposes, this type of roaming can be considered the same as the first type.

c. *Inter-ESS:* Handoffs between different APs from different network providers. In this type of roaming, the IP address of the phone will need to change. Also, the security used in the association will be different. We will discuss this type of roaming, briefly, at the end of the chapter.

d. *Inter-Network*: Handoffs between 802.11 and other wireless networks such as CDMA or GSM. This is the "holy grail" of 802.11 roaming, and requires call-signaling and network-infrastructure support. This type will be discussed in Chapter 10.

8.3 Roaming Issues

For 802.11, roaming issues may be classified into three categories:

i. *When to Roam*: Determining that the STA indeed needs to move from its current associated AP to another AP.

ii. *Where to Roam*: Determining which AP to move-to / associate-with next.

iii. *How to Roam*: Defining the exact process of disconnecting from an AP and connecting to another AP.

We first look at how the base 802.11 protocol addresses these issues.

8.3.1 Basic 802.11 Roaming Support

There are no explicit roaming procedures/commands in the base 802.11 standard. However, the base standard supports Intra-ESS[2] roaming implicitly through the association/deassociation procedures. A STA can roam though multiple APs in an ESS by using basic 802.11 operations as described below:

a. STA connects to/associates with the strongest[3] AP in the BSS.

b. STA stays connected to the AP until disassociated.

c. When disassociated, STA scans all channels to find the strongest AP available.

d. STA authenticates and associates with the new AP.[4,5]

---

[2] Types (a) and (b) as described in section 8.2.
[3] Strongest usually in the sense of RSSI (Received Signal Strength Indication).
[4] If 802.11i/WPA/WPA2 is in use, the STA re-establishes 802.11 security as part of the authentication process.
[5] If 802.11e/WMM is in use, STA re-establishes QOS services if necessary.

8.3.1.1 Basic 802.11—How to Roam

Given this process, let's examine how the basic 802.11 standard addresses the three roaming issues. We first examine the question of how to roam. When a STA wishes to connect to an 802.11 network, it broadcasts a Probe Request message on all channels. This message contains the SSID of the network (ESS) that this station wishes to connect to.[6] An AP that receives the Probe Request message may reply back with a Probe Response message if it wants to allow this station to connect to its network. On receiving the Probe Response message, the STA starts the authentication process. The aim of the authentication process is to establish reliably the identities of the communicating stations. The next step in the process is association. The aim of the association process is to establish a logical connection between the STA and the access point. The association process starts with the STA sending an Association Request to the access point. This request contains parameters such as the capability info and the rates that the STA can support. The AP responds with an Association Response message, which may accept or reject the association depending on the parameters provided in the Association Request message. Once a STA is associated with the access point (and therefore the BSS), the network now knows "the location" of this STA and can deliver data destined for the STA. The interesting thing to note here is that the process of a STA connecting to an AP is the same irrespective of whether or not the STA was previously connected to another AP. Hence, in the basic 802.11 standard, the roaming process becomes just another instance of connecting to an AP.

8.3.1.2 Basic 802.11—When to Roam

From the above process, it is important to note that the decision of when to roam is not explicitly addressed and is left to the implementation. Most implementations rely on some combination of the following approaches to determine when to disassociate from the current AP:

- AP explicitly sends a Disassociation message to the STA

- STA fails to receive a threshold number of beacons from the AP

- STA fails to transmit a threshold number of packets to the AP/network.

- STA sees its or the AP rate-adaptation algorithm locking in on the lower rates

8.3.1.3 Basic 802.11—How to Roam

Also, note that, even though the standard defines the process by which a STA may discover available APs (scanning), the decision of where to roam —i.e., which AP to connect to from a list of available APs—is not explicitly addressed in the standard and this decision is left to the

---

[6] The SSID field may be left blank in order to indicate that the station wishes to connect to any network that it finds.

implementation. Most implementations rely on the signal strength, aka RSSI (received signal strength indication), to decide which AP to connect to.

There are two kinds of scanning methods defined in the standard: passive and active. In the passive mode, the STA waits and listens for beacons, which are transmitted periodically by APs. In the active mode, apart from listening for beacon messages, the STA sends additional Probe-Request broadcast packets on each channel and waits to receive responses from APs. Thus, in active scanning, the station actively probes for the APs.

Now that we have seen how the basic 802.11 standard handles the three roaming issues of when, where and how to roam, we look at why these standard methodologies are insufficient for implementing VoWLAN and how this problem can be solved.

8.4 Roaming and Voice

The basic 802.11 approach works well for data applications because most such applications use a reliable transport layer protocol like TCP which conceals the delay/packet loss due to the handoff by using retransmissions. In other words, higher layers (HTTP, the web browser and the user) are unaware of 802.11 handoff and can continue without disruption. Even higher-layer applications that use an unreliable transport layer protocol like UDP, would "see" the handoff only as temporarily increased delay or packet loss. As long as the application can tolerate and recover from this delay/loss, an 802.11 handoff would be "transparent" to the user in this case too.

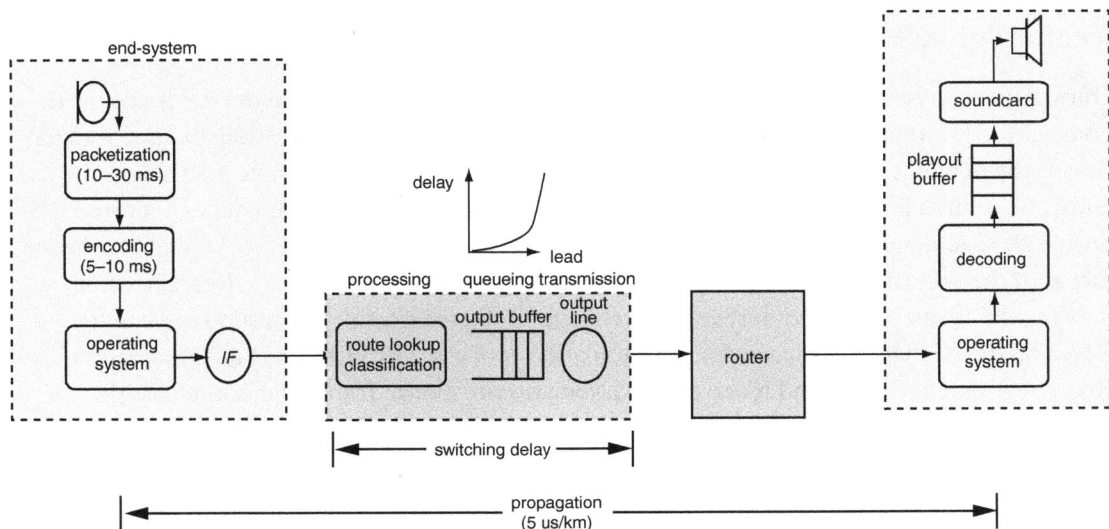

Figure 8.2: Delay Components in the VOIP Voice Path

As explained in section 5.8, the challenge for VoWLAN is that voice is extremely sensitive to delay. The end-to-end delay budget for voice is 250 ms; this means that the accumulative delay between the two endpoints involved in a voice call must not exceed 250 ms. This 250 ms must include the total transmission delay, propagation delay, processing delays in the network and codec delays at both endpoints. In WLANs, the 802.11 MAC introduces an extra transmission delay in the WLAN due to the collision and backoff nature of the MAC protocol.[7] This wireless transmission delay increases as networks become more congested or suffer from interference. The bottom line is that the budget for each component of the accumulative delay will typically be specified by the service provider, but we can't assume that we would have all 250 ms available for 802.11 delays. A good VoWLAN implementation would aim to keep the 802.11 delays limited to 40–50 ms. However, basic 802.11 handoff times have been shown to be on the order of a few hundred milliseconds[8] and are therefore unacceptable for voice. Therefore, the challenge for VoWLAN is to reduce the handoff times involved in roaming so that voice packets suffer only minimal delay.

Mishra et al.[9] divide the handoff process into two distinct logical steps: discovery and re-authentication. Attributing to mobility, the signal strength and the signal-to-noise ratio of the signal from a station's current AP might degrade and cause it to lose connectivity and to initiate a handoff. At this point, the client needs to find the potential APs (in range) to associate to. This is the discovery phase of handoff and is accomplished by scanning. During the scan process, the STA either waits for beacon messages (passive scanning) or sends out Probe Requests and waits for Probe Responses (active scanning) on available channels. The latency introduced during this process is known as the Probe delay. Probe delay is the dominating component in the overall handoff delay. As Mishra et al. have shown, probe delay accounts for more than 90% of the overall handoff delay.

During the reauthentication phase, the STA attempts to reauthenticate to the AP it wishes to connect to. The reauthentication process typically involves an authentication and a reassociation to the new AP. The delays incurred during this process are divided into two categories: authentication delay and reassociation delay. Authentication delay is the latency incurred during the exchange of the authentication frames between the STA and the AP and is a strong factor of the security protocol being used (WEP, WPA, WPA2, CCX, etc.). Reassociation delay is the latency incurred during the exchange of the reassociation frames between the STA and the AP. Upon a successful authentication process, the STA sends an Association-Request frame to the AP and receives a reassociation-response frame and completes the handoff. Future implementations may include additional IAPP (Inter Access Point Protocol) messages during this phase, which will further increase the reassociation delay.

---

[7] The situation is worse if both endpoints in the voice call use VoWLAN.
[8] Typically between 200 and 500 ms.
[9] *An Empirical Analysis of the IEEE 802.11 MAC Layer Handoff Process*, Mishra, Shin, Arbaugh.

With respect to packet loss, Voice-over-IP codecs can typically handle 1–3 % *uniform* packet loss without too much call-quality degradation. This is illustrated in Figure 8.3, which graphs a perceptual evaluation of voice quality (PESQ) score versus packet-loss percentage for various codecs (PESQ is a standard method of measuring voice quality). A PESQ score of 4 or above is equivalent to toll—i.e., PSTN—call quality. A score of less than 3.7 implies that there are noticeable impairments. Unfortunately, the packet-loss concealment techniques (discussed in Chapter 3) cannot handle a large burst of lost packets. This is exactly what will happen in a roaming situation.

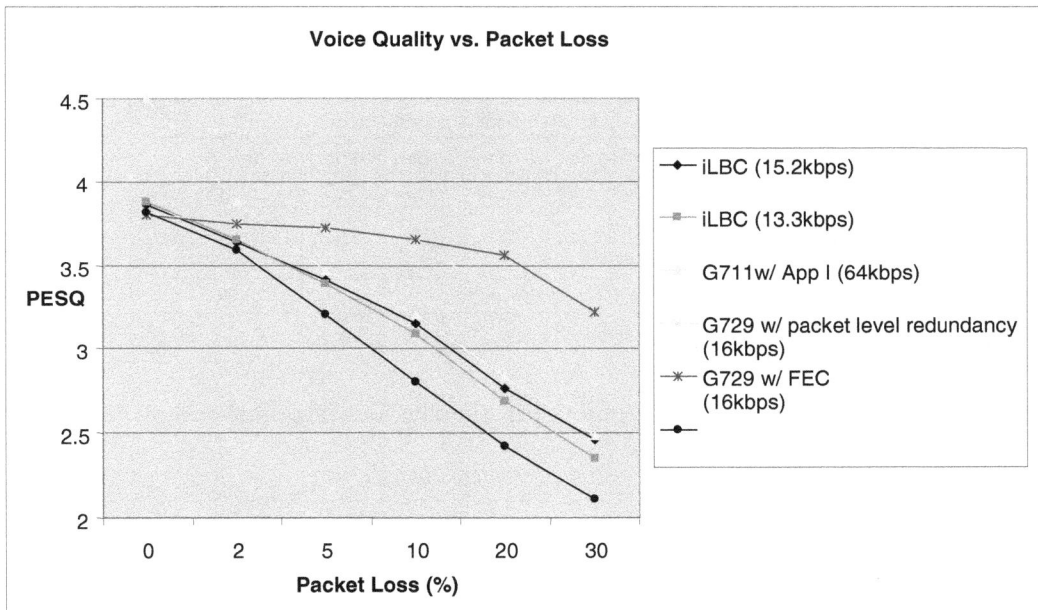

Figure 8.3: Voice Quality vs. Packet Loss Rate

Thus, we can conclude that a more sophisticated approach is required for this problem. Such a comprehensive approach should focus on the reduction of probe, authentication and reassociation delays. We examine some techniques for such an approach in more detail in the following sections.

8.5 Preparing to Roam: Scanning

Scanning is the act of searching 802.11 channels for available APs. The aim and end result of scanning is to maintain a list of available candidate APs that the STA can connect to, if it is disconnected, or that are candidates to connect to if the current AP connection deteriorates. This list, often referred to as the *Site Table*, is a data structure in the Voice-over-Wi-Fi phone's WLAN subsystem. It maintains, for each AP that is known, information such as the AP channel and the RSSI. Table 8.1 lists typical Site Table data elements.

Table 8.1: Typical Site Table Elements

Data Element	Description
BSSID	AP Mac Address
SSID	AP SSID
Last Packet Seen	Time stamp of last packet seen
RSSI	Signal strength of last packet seen
Beacon period	Time between consecutive beacons
DTIM period	Time between consecutive beacons carrying Broadcast Indication Maps
TSF	Timestamp when beacon is expected
Last Associated	Timestamp of when we last associated with this AP
Last Disassociated	Timestamp of when we last disassociated with this AP
Security Parameters	If preauthenticated
Channel	802.11 channel
Loading	802.11 QOS loading factor
Capabilities	PHY, QOS, etc.

To facilitate roaming algorithms, it is helpful to partition the Site Table into several categories:

- The last AP that we associated with prior to the current one.

- Previously visited APs (APs that we have associated with in the past and can revisit if necessary).

- APs that we can roam to, but have not visited yet.

- APs that we have preauthenticated with.

8.5.1 Scanning Types

Given the central importance of probe delays (aka discovery times) in reducing handoff times, it is not surprising that the scanning process forms an important product differentiator in the VoWLAN market. A good VoWLAN scan algorithm will attempt to balance maintenance (updating the status of known APs) and discovery (locating new APs) scanning, keeping in mind that each scan operation will drain precious battery power. We discuss here some of the choices available for designing a good VoWLAN scan algorithm.

8.5.1.1 Passive or Active

In passive scanning, the STA will tune to an 802.11 channel, turn on its receiver and wait for AP beacons to be received. This strategy is power intensive since the phone will need to keep the receiver powered up for at least one beacon period for each channel (i.e., 100 ms typically). Unfortunately, there may be times where it is the only allowed technique due to regulatory issues, as in 802.11 a/h networks where regulatory constraints prohibit transmission on a new channel until 802.11 traffic is first "seen." This is to prevent interference with other spectrum users such as radars. Passive scanning is parameterized with the channels to scan, the maximum amount of time to listen, and the maximum number of AP beacons and Probe Responses to collect.

In active scanning, the STA actively participates in the scanning process by transmitting Probe-Pequest messages on different 802.11 channels to elicit a Probe Response from available APs. This strategy is less power intensive than passive scanning as long as the response/turnaround time of an AP is less than the beacon time. Since this is usually the case, the STA will need to keep the receiver powered up for a much shorter time, thus saving power.

8.5.1.2 Background or Foreground

Background scanning takes place while the STA is still connected to an AP. One aim of this type of scanning is to update the Site Table for known, neighbor APs that might be good candidates to roam to, should the need arise. This is also referred to as Site Table maintenance. The goals of Site Table maintenance are to (a) ensure that the APs in the Site Table are still reachable, and (b) determine the ordering of the AP in case roaming becomes necessary. A second aim of background scanning is to learn about APs that may now be in range because of mobility. This is also referred to as *AP discovery*. In either case, background scanning will make heavy use of 802.11 doze mode to be able to stay connected to the primary AP (and still receive packets) and yet be able to perform operations with other APs on other channels. This is the only way that a vanilla 802.11 b/g station can ensure that it won't lose data when temporarily listening on another channel.

Given that we are trying to reduce the handoff delays, a step in this direction is to start the discovery process before the STA actually loses communication with the current AP. Hence, background scanning can serve to reduce handoff times by reducing probe delay. Taking this idea to the extreme, a STA which continues to scan for available APs at all times would achieve the smallest handoff times. However, continuous background scanning is not an optimal approach since scanning consumes power and most 802.11 STAs with the need to roam are battery-operated devices where power consumption must be minimized to achieve long battery life. A balance is therefore needed in the background-scanning frequency. An often-used approach is to start frequent background scanning once threshold levels in certain metrics (such as received AP beacon RSSI) are reached. This allows the STA to start the background scanning (i.e., the discovery) process before risking total disconnection with the current API, but not when it is too far away[10] from new APs (and thus can see good roaming candidates). Well-designed background-scanning algorithms minimize the need for foreground scanning (thus reducing handoff times) without consuming too much power.

Foreground scanning takes place when the phone is not in an associated state. The aim is to find an AP to connect to as soon as possible. Regardless of how well-designed the background scan algorithm is, foreground scanning is sometimes necessary—for example, due to mobility (the STA may move into an area of no coverage and then come back) or due to the idiosyncrasies of the wireless medium (sudden drops in RSSI).

8.5.1.3 Broadcast or Unicast

With *broadcast scanning*, the STA issues a Probe Request with a broadcast address and then waits for some number of Probe Responses. Alternatively, if used with passive methodology, the STA waits for some number of beacons. This technique is useful for AP discovery. This scanning is parameterized with the channels to scan, the SSID of interest, the number of Probe Requests to issue, the amount of time to wait for a response, and the maximum amount of time to listen on each channel.

With *unicast scanning*, the Probe Request contains a specific BSSID. This technique is appropriate for Site Table maintenance since it requires that the AP already be discovered (i.e., BSSID known). This technique is parameterized by the 802.11 channels to scan, the SSID and BSSID to scan, the number of Probe Requests to issue, and the retry interval.

8.5.1.4 Consolidated and Sliced Scanning

With *consolidated scanning*, all channels that need to be scanned are scanned one after the other—i.e., as soon as the scan of one channel is complete, the scan of the next channel is started. The reasoning behind the approach is to complete the scanning process as soon as possible. While appropriate in most circumstances, this approach creates a problem when the STA is involved in a voice call.

Since voice packets are transmitted periodically (typically once every 20 ms) and since a consolidated scan may take as long as a few hundred milliseconds, a *sliced scanning* approach will be preferrable. In this technique, the scan is broken into small chunks, with each chunk being timed to take place between voice packets. This technique is useful for Scan Table maintenance or AP discovery while the STA is involved in a voice call and has the advantage that the scan operation's impact on voice is minimized. Although sliced scanning can be combined with both passive and active scanning, the long delays in passive scanning usually defeat the purpose of sliced scanning and therefore sliced scanning is typically used only with active scanning.

---

[10] RSSI is a good indicator of distance between the AP and the STA.

8.5.1.5 *Synchronous or Asynchronous Scanning*

When to start scanning is typically a decision of the scan algorithm. When the start of a scan is independent of the AP timing, it is referred to as *asynchronous scanning*. On the other hand, *synchronous scanning* (also called *timed scanning*) is a passive scan that uses previously saved information on the APs in the Scan Table to time when to turn on the radio in order to receive a beacon from the desired AP. Thus, no Probe Request needs to be sent, and the phone radio can be turned on just for the beacon reception. This is the most power-efficient scanning technique and is applicable for Site Table maintenance. However, this technique has limitations in that, over time, clock drift between each AP and the phone may cause the phone to wake up too early or late and miss the beacon. Therefore, this technique needs to be used in conjunction with the other methods.

8.5.2 *Scanning Strategies*

Coming up with a good scanning strategy for a VoWLAN STA is a definite art form and we should expect that phones scanning and roaming algorithms will be key differentiators from vendor to vendor. The basic idea is to balance keeping the AP roaming candidate list up to date with phone power and performance requirements. During disconnected scanning, the goal is to find a suitable AP without draining the battery too quickly. The power impact of these various scanning types is illustrated in Figure 8.3. A state-of-the art implementation will dynamically adjust its scanning approach, depending on the battery level. For example, with low battery, the phone may forgo or reduce the amount of background scanning when the phone is not in use.

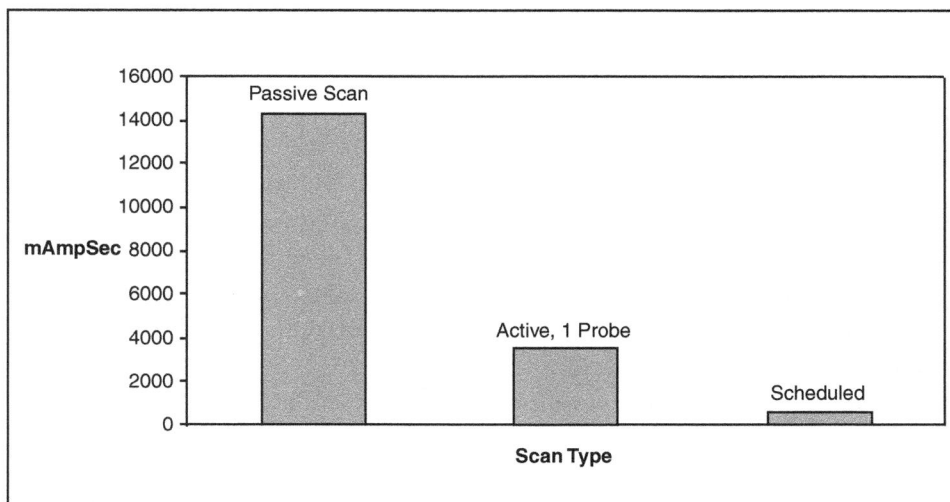

Figure 8.4: Current Consumption vs. Scan Type

8.5.3 Other Site-Table Management Techniques

Given that the discovery process is the most time-intensive process in handoffs, any handoff scheme that uses techniques/heuristics that either cache or deduce AP information without having to actually perform a complete scan clearly stand to benefit from the dominating cost of the scan process.

8.5.3.1 AP Neighbor List

One powerful technique is the use of AP neighbor lists, as proposed in the 802.11k standard,[11] and some proprietary 802.11 extentions. With this technique, each AP is responsible for sending down the list of "suitable" neighbor APs[12] (their BSSIDs and channel numbers) in their beacons and Probe Responses. This list essentially removes or at least lessens the need for discovery scanning by the STA.

Various approaches are possible for an AP to determine the neighbor list. One method is to provision each AP manually. The challenge with this technique is reliability—i.e., how accurate is the list provided by each AP. Furthermore, as network topology changes, updating each AP with the correct neighbors may be error prone. Another approach, that taken by 802.11k, is to have the STAs effectively provide information that the AP can use to build the list through what is known as radio measurements. We will discuss 802.11k and roaming below.

8.5.3.2 Channel List

Another area where the AP can provide assistance is to provide a list of channels to scan. While 802.11 b/g deployments are limited to at most 14 channels, 802.11a can exist in environments with up to 20 radio channels. Clearly, not all channels will be used in one facility. If the STA knows which channels to scan, it can simplify its scanning operation, reduce scanning times and save power.

8.5.3.3 Site-Table Pruning

Site Table management will need to include an *aging* or *pruning* algorithm that can remove APs that are no longer available or are no longer good candidates for roaming to. Over time, the number of APs that are discovered will grow. To avoid filling up the Site Table and to minimize the list of candidates, the Site Table manager will need to periodically remove APs from the list. The trick here is to decide what criteria to use. Some options include age (i.e., drop the AP that the phone has not associated with for the longest time), signal strength (i.e., drop the weakest), etc. Again, this is an area where a lot of product differentiation can be expected.

---

[11] This approach is already present in some proprietary AP messages/information elements.
[12] To which the STA can roam, if needed.

8.6 When to Roam

A good scanning algorithm is a strong strategy for reducing handoff times. Another important strategy is not to wait for the STA to be disconnected from the current AP before starting the actual handover. Recall from section 8.3 that the base 802.11 standard uses a "brute force" approach to determine that disassociation has occurred. In other words, a STA waits until some number of beacons are missed or lost or until a threshold limit on packet retransmission is reached to start discovery of other available APs. Adopting this approach means that the STA must necessarily lose some packets before a handoff is initiated. However, there is nothing in the 802.11 standard to prevent the STA from disconnecting from its current AP and connecting to another AP whenever it chooses to do so. In other words, the STA can handoff from one AP to another at any time it deems appropriate. A simple improvement in this regard is for the STA to initiate the handoff when the running-average-RSSI falls below a certain threshold. Note that this threshold might vary depending on whether or not the STA is actively involved in a voice call.

It is important to distinguish *instantaneous RSSI* from *running-average-RSSI*. In this book RSSI refers by default to running-average-RSSI. The former refers to the RSSI of the last packet observed from the AP, while the latter refers to the average RSSI as calculated over the last N (>1) packets received from the AP. To understand the importance of running-average-RSSI, it is important to realize that, in wireless media, the instantaneous signal strength at a given distance is affected by many factors. One of the most important considerations that determines the instantaneous signal strength is, not surprisingly, the operating environment. For example, rural areas with smooth and uniform terrain are much more conductive to radio waves than the more uneven (think tall buildings) and varying (moving automobiles, people and so on) urban environment. The effect of the operating environment on radio propagation is referred to as *shadow fading* (*slow fading*). The term refers to changes in the signal strength occurring due to changes in the operating environment. As an example, consider a receiver operating in an urban environment. The path from the transmitter to the sender may change drastically as the receiver moves over a range of a few feet. This can happen if, for example, the receiver's movement resulted in the removal (or introduction) of an obstruction (a tall building perhaps) in the path between the transmitter and the receiver. Shadow fading causes the instantaneous received signal strength to be lesser than (or greater than) the average received signal strength.

Another factor that strongly affects radio propagation is *Raleigh fading* (*fast fading*). Unlike slow fading, which affects radio propagation when the distance between the transmitter and the receiver changes on the order of a few feet, fast fading describes the changes in signal strength due to the relative motion of the order of a few centimeters. To understand how such a small change in the relative distance may affect the quality of the signal, realize that radio waves (like other waves) undergo wave phenomena like diffraction and interference. These phenomena lead to multipath effects; in other words, a signal from the transmitter may reach the receiver from multiple paths. These multiple signals then interfere with each other at the receiver. Since this interference can be either constructive or destructive, these signals may either reinforce each other or cancel each other out. Whether the interference is constructive or destructive depends on the path length (length the signal has traveled) and a small change in the path length can change the interference from a constructive to a destructive one (or vice versa). Thus, if either the transmitter or the receiver move even a few centimeters relative to each other, this changes the interference pattern of the various waves arriving at the receiver from different paths and a constructive interference pattern can be replaced by a destructive one (or vice versa). This fading is a severe challenge in the wireless medium since it implies that, even when the average signal strength at the receiver is high, there are instances when the signal strength may drop dramatically.

Another effect of multipath is *inter-symbol interference*. Since the multiple paths that the signal takes between the transmitter and the receiver have different path lengths, this means that the arrival times between the multiple signals traveling on the multiple paths can be on the order of tens of microseconds. If the path difference exceeds 1 bit (symbol) period, symbols may interfere with each other and this can result in severe distortion of the received signal.

The bottom line is that roaming decisions should be based on running-average RSSI rather than instantaneous RSSI, since this approach prevents having too many handoffs.

Besides using the RSSI as an indicator for deciding when to roam, a roaming algorithm will also need to consider the following exception events for triggering a roam decision:

- Missed beacons: If the phone misses "N" beacons from the current AP, this is a good indication that it is time to roam.

- Consecutive packet retransmission: If the phone suddenly has to retransmit every packet to the AP, then it is time to roam.

- Disconnect message from the current AP: No discussion here. We need to roam now!

- Security handshake/re-key failures.

8.7 Where to Roam

Once the phone decides that it needs to roam to another AP, the next step is to determine which AP the station should roam to. Assuming that the scan algorithm has been well designed, the STA will have access to an up-to-date Site Table, and this decision is reduced to parsing the Site Table and choosing an appropriate AP. We expect that this too will be an area of differentiation between implementations. The basic idea of selecting an AP from the

Site Table will be to compare the "quality" of the current AP link to the predicted "quality" of roaming candidates. *Link quality* is typically measured by RSSI. As defined earlier, RSSI gives a measure of signal strength in wireless networks. Each entry in the Site Table will have a corresponding RSSI and a basic roaming algorithm will compare the RSSI of the available APs in the Site Table to decide where to roam to.

A more sophisticated roaming algorithm might consider the following:

- Skip over the last AP with which we had last associated, at least for a time period following the last roam. This will avoid a situation where the phone might flip-flop between two APs.

- Skip over APs that we have not preauthenticated.

- Factor in QoS loading, if available, so as to load balance across APs.

8.8 Reauthentication Delays

Section 8.4 discussed the various delays involved in a handoff process. Since probe delays account for the bulk of delay in a basic 802.11 handoff, most vendors concentrate on reducing the probe delays by improving their scanning algorithms. However, it is also important to concentrate on reducing reauthentication delays to reduce the overall handoff delay.

Recall from Chapter 4 that the 802.11 standard does not allow a STA to be associated simultaneously with two different APs. For this reason make-before-break[13] handoffs (as in GSM, UMTS, etc.) are not possible within the 802.11 standard. Hence, the STA cannot start its association process with the new AP unless it has disassociated from its old AP. However, the 802.11 standard does not prevent STAs from being authenticated with multiple APs. This gives rise to the concept of preauthentication, wherein a STA while connected to its current AP may authenticate itself with other APs *through the current AP* so that, if and when a handover is carried out, the handover delay can be minimized.

Preauthentication allows a phone to perform WPA2 authentication (e.g., EAP-TLS) with APs that it discovers during its background scanning. This authentication is tunneled through the AP with which it is currently associated and the resulting pairwise keys are cached in the Site Table. The advantage of this technique is that, when a phone determines that it is time to roam, the candidate AP list can be pruned to those that have been preauthenticated. Furthermore, the handover procedure is shortened because EAP messaging is reduced.

---

[13] The STA/Mobile Phone associates with the new AP/Base-Station before disassociating with the current AP/Base-station.

8.9 Inter-ESS Roaming

In inter-ESS roaming, the VoWi-Fi phone can roam onto an entirely different distribution system. There are other issues with this type of roaming that are more severe than just the level-2 Wi-Fi roaming time. Roaming in this environment will involve level-3 (and above) actions as well because the phone's IP address will most likely need to change. With an IP address change comes IP address-assignment delays (DHCP handshaking), and call-signaling reinitialization (e.g., SIP registration). With this type of roaming, simply keeping an active call alive is a big challenge.

A solution for this type of roaming will involve a protocol known as *Mobile IP* (RFC 2002). The key elements of mobile IP are illustrated in Figure 8.5.

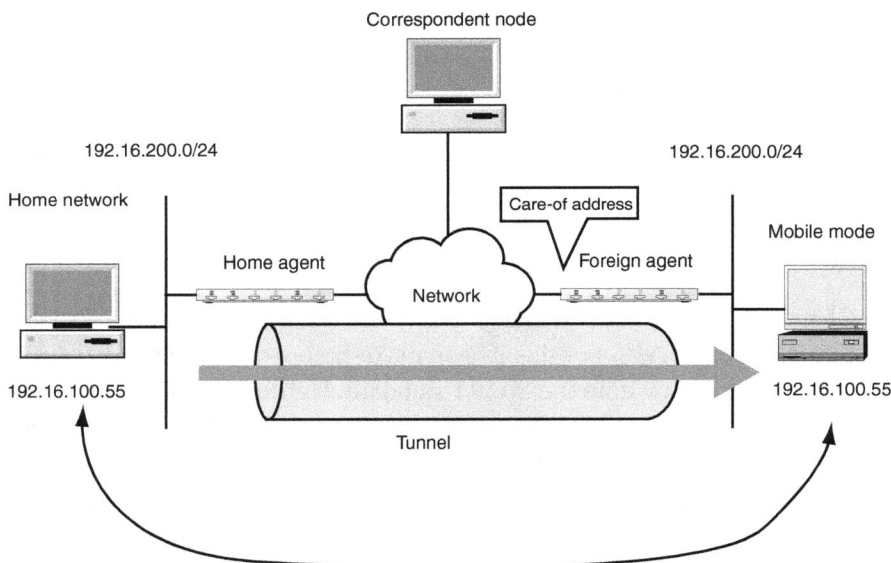

Figure 8.5: Mobile IP Example

The first element is the mobile node, in our case the VoWi-Fi phone. This element has a relationship with a mobile IP element known as the home agent. The home agent is typically implemented in a router in the mobile element's home network, but the relationship is maintained even if the mobile node moves to another network. The home agent is responsible for providing a location-invariant representation of the mobile node (i.e., a fixed IP address) to the network. This element advertises itself as the destination for traffic being routed to the mobile node, and then forwards it to the mobile node's current location (known as the care-of address). This is done via an IP tunnel, possibly secure.

The final element in mobile IP is the foreign agent. This element exists in networks to which the mobile node may roam (again typically as a component of the local router), and is responsible for updating the home agent that the mobile node is now located on its network. It can be involved in the IP tunnel back to the home agent.

Consider the example of a VoWi-Fi phone that is mobile-IP capable and has a call established and is currently located in its home network. In this case it will receive VoIP call signaling and media traffic directly as in an normal VoIP case. When this phone roams onto a new ESS, it will obtain a new [local] IP address (e.g., through DHCP). At this point, it will determine the address of the mobile-IP foreign agent on this network and attach itself to it via registration. The standard defines an agent-discovery protocol using advertisement solicitation messages from the mobile nodes and advertisement messages from the agents.

The registration message contains the mobile node's current care-of address and where the home agent is located. The foreign agent will communicate the new location (care-of IP address) of the phone back to the home agent. The foreign agent can establish an IP tunnel to the home agent, or can facilitate the tunnel formation directly to the phone. In either case, the home agent will now have a path to route received packets (e.g., SIP signaling or media) to the phone.

The home agent configures itself to intercept IP packets and reroute them to the mobile node at its care-of address. To the call servers or proxies and the destination of the call, our phone is still using the original IP address. Packets from the mobile phone can either be sent directly from the phone, or in some cases, forwarded to the home agent first.

While mobile IP will allow inter-ESS roaming, it will in general not be seamless. In addition to the level-2 roaming delays described above, inter-ESS will incur the following level-3 delays:

- Acquisition of a new, care-of, IP address

- Foreign agent discovery

- Registration with the home agent—this step may involve authentication

- IP tunnel setup, including possible security association negotiation.

8.10 Future Enhancements

8.10.1 802.11k

802.11k is a proposed extension to 802.11 that adds radio measurement and radio management features. At the time of writing this book, 802.11k is still in draft form. We have touched briefly upon its impact on roaming and scanning, in particular two reports that are provided by the AP:

- Channel report

- Neighbor report

The *channel report* can be sent by the AP in Association and Probe Responses. It contains a list of channels where a phone is likely to find APs. There can be one report for B/G and one for A channels, if the deployment has fielded mixed B/G and A networks. A phone can use this report to restrict the channels that it scans, thus speeding up the Site Table maintenance activities, and saving power.

The *neighbor report* can be sent by the AP in associations and Probe Responses. There is also a mechanism for the STA to explicitly ask for the report (neighbor report request). The report contains of a list of neighbor (i.e., locally available) APs. The following information is included for each AP in the list:

- BSSID

- Channel number

- Regulatory class (i.e., B/G or A)

- PHY options

- Capabilities (as would be seen in the AP's beacon)

- TSF (optional): This gives the offset from the current AP's beacon when this candidate AP sends its beacon. There is a note in the specification that this information should only be sent if the accuracy is within a certain tolerance.

- Reachability bits: A 2-bit flag is included to indicate if this AP can be reached from the current AP for purposes of preauthentication.

- Security flag: An indicator of whether the security options for this AP are the same as the current AP.

This information fits nicely into what is required for a good Site Table and if it is 100% accurate, can remove or reduce the need to perform discovery scanning. Maintenance scanning will still be required, however, since the report can't give any link-quality information for candidates. The specification does not define where the AP is to obtain the information in the report. As discussed above, provisioning each AP with the information is an option, but the most likely source of the information will be from radio measurements reports provided by the stations to the AP. There are various types of station-provided measurement reports defined in the 802.11k standard, including:

- Channel-load report (giving the percentage of time the station sees a channel as busy).

- Channel-noise histogram.

- Beacon report (this is a report generated by the station about a particular BSSID—for example, the received power of the beacon or Probe Response, and a copy of the capability information field).

- Hidden-node report.

- Station statistics (for example, frame received/transmitted and exception-condition counts).

8.10.2 802.11r

802.11r is a proposed extension to 802.11 to deal with roaming issues such as those discussed previously. Specifically, the purpose of 802.11r is to "enhance the 802.11 medium access control (MAC) layer to minimize the amount of time data connectivity between the station (STA) and the distribution system (DS) is lost during a Basic Service Set (BSS) transition." At the time of this book, this standard is in a great deal of flux, with at least two competing proposals. Among the topics to be addressed by this standard are:

- Security, in particular techniques to reduce the 802.11i handshake latency by introducing a new key management hierarchy.

- QoS, in particular ways to signal and negotiate QoS requirements early on in the roaming process.

8.11 Conclusion

In this chapter we have looked at some of the issues with roaming in Wi-Fi networks. We discussed the reasons for roaming and some of the techniques that are used. Achieving seamless mobility, where a phone user does not experience a noticeable loss of service during the roaming process, is possible with a carefully tuned scanning and roaming algorithm. This algorithm will be a key differentiator in VoWi-Fi products and must balance multiple factors including battery life. We presented a brief overview of mobile IP, an internet protocol that can be used for more complicated roaming scenarios. We concluded with some of the work-in-progress in the 802.11 standard that is addressing this issue.

Power Management

9.1 The Need for Power Management

Almost every handheld device today faces the challenge of maximizing battery lifetimes. One approach to achieve this is to increase battery capacity. However, this approach is limited by increasing size and weight of the batteries as their capacity increases. Given the small size of these handheld devices, there is a limit on how big the batteries can grow. Therefore, a parallel or alternative approach is to conserve the available power in order to increase battery lifetimes. This is known as power management and is the topic of this chapter.

As we said, power management is an issue that is common to most handheld devices. However, the wireless domain makes the issue even more significant. The radio modules used for wireless transmission and reception are extremely power hungry. If appropriate power-management techniques are not implemented, the radio interface will drain the battery of a Wi-Fi phone[1] very quickly, making it practically unusable. As we shall see in this chapter, a good power-management scheme requires a systems solution to be successful.

9.2 Underlying Philosophy of Power Management

To implement a power-management strategy in order to minimize power consumption, it is first required to understand where the power is consumed in Wi-Fi phones. Most Wi-Fi handsets would typically use a system-on-chip (SoC) device as the host CPU. This means that the "host" shown in Figure 9.1 would consist of various on-chip peripherals besides a micro-processor. The term "host processor" is therefore used in this chapter to refer to the complete SoC, which includes the microprocessor and the on-chip peripherals. Figure 9.1 shows a high-level schematic of a Wi-Fi phone.

---

[1] Note that, since the AP is not a handheld/mobile product, it does not need power management.

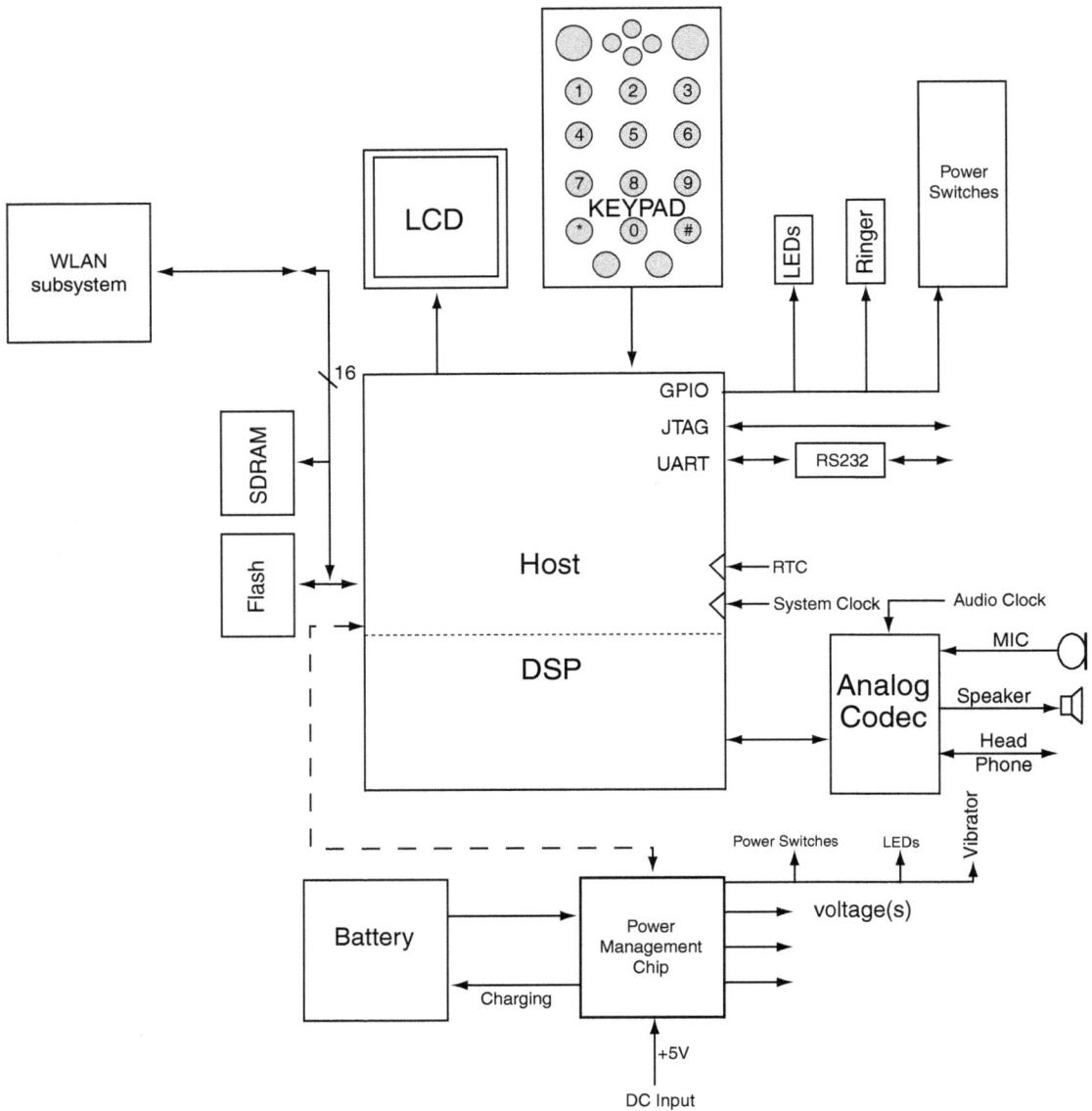

Figure 9.1: Typical Block Diagram of a Wi-Fi Phone

Our analysis has shown that the major power consumers in a Wi-Fi phone (like the one shown in Figure 9.1) are:

- Radio/WLAN subsystem (see section 9.4.1)

- Host processor (see section 9.4.2)

- LCD (see section 9.4.3)

- Backlight (see section 9.4.3)

- DSP (see section 9.4.4)

- Analog codec (see section 9.4.4)

- Flash, SDRAM (see section 9.4.5)

- LEDs (see section 9.4.6)

- For SoCs, on-chip peripherals (see section 9.4.6)

- On-board peripherals (see section 9.4.6)

- Etc., i.e., drainage current (see section 9.4.7)

The list above is (more or less) in descending order of power consumption. That is, the radio/WLAN subsystem usually consumes the most power, followed by the host processor, the LCD, backlight, etc. Obviously, the exact ordering of such a list is product specific and depends on the choice of components. However, the underlying concept of power management does not change.

The underlying philosophy of power management is that a system does not use all its components/resources all the time. This implies that at any given time, components that are not being used can be temporarily shut off to reduce the power consumption of the system. An example of such a component is the LCD on a wireless handset, which can be switched off when the phone is not being used. An obvious assumption of this approach is that the said component can be turned back on when required within acceptable time limits.

This binary philosophy of power management, where unused components are shut off and components being used remained powered on, can be extrapolated so that components that remain powered on are controlled dynamically such that their power consumption is in proportion to their load (amount of work they are doing). An example of such an approach is the dynamic control of the clock frequency of a processor, depending on what the processor is doing. In summary, the philosophy of power management is to conserve power by using only what is necessary only when it is necessary.

9.3 Designing for Power Management

While designing any system, an architect has to keep in mind various criteria for selecting the hardware and software components. For a handheld design like a Wi-Fi phone, power-consumption and power-management features should form an important criterion for system design. In this section, we first look at how power-management considerations are involved at the system-design level. Then, we dig a little deeper and see how power considerations affect hardware design. Finally, we look at what can be done in software design to maximize efficient usage of power.

9.3.1 Power-Aware System Design

9.3.1.1 Selection of Hardware Components

Selection of hardware and software components is one of the most challenging tasks of an architect. Since this stage occurs very early on in the project, usually with limited information about the field requirements and future evolution of the product, the choice is fraught with risks. Adding power considerations to the matrix makes the choice a little easier by narrowing down the choice. On the other hand, it also complicates the process since the architect now has to take power consumption into account while evaluating the trade-offs. This section talks about some system-design considerations which, in our experience, have significant impact on the power consumption of a Wi-Fi phone.

As pointed out in section 9.2, the WLAN subsystem is one of the biggest power consumers. The design of this subsystem is therefore of utmost importance. While selecting the chipset to be used for the radio and baseband processing for 802.11, power should be one of the top criteria. The architect should look not only at the min/max and average power consumption numbers in active and standby mode, but should also take into account the timing requirements of transitioning from one mode to another. The timing requirements will define the minimum time for which the subsystem can be put in standby mode, which in turn determines how often the subsystem can be put in standby. Some chipsets may provide multiple standby modes. In such scenarios, it is essential to understand the differences between these modes, the corresponding power-consumption numbers, and the timing involved. Such a detailed study, along with the system requirements, would be required to determine which mode would actually be used by the Wi-Fi phone.

The host processor is another important power consumer in the overall system. Therefore, choice of the host processor has significant impact on battery life. As a yardstick, note[2] that an Intel 386 processor consumed about 2 watts of energy, whereas a Pentium 4 can use as much as 55 watts. Again, this choice is not as simple as it seems. Besides looking at the minimum, maximum and average power-consumption numbers, the architect must look at how the power savings are obtained.

---

[2] From "Automatic Performance Setting for Dynamic Voltage Scaling,"Flautner et al.

In general,

$$P_{AC} = C_{eff}V^2_{dd}\ f$$

where,

P_{AC} is the dynamic power consumption

V_{dd} is the supply voltage

C_{eff} is the average switched capacitance per cycle and

f is the clock frequency.

Many processors use frequency scaling to control the power consumption. This technique involves lowering the operating frequency of the processor to reduce the power consumption. However, this reduction in the operating frequency comes at the cost of a "slower" processing capability and tasks take longer to complete. Therefore, the architect must take into account when and how often this capability can be utilized. Some processors also allow the use of voltage scaling, which involves lowering the operating voltage of the processor. Dynamic voltage scaling is an especially powerful technique since, as shown in the equation, the power consumed in a processor varies as the square of the applied voltage. However, voltage scaling may come with certain constraints attached—for example, lowering the operating voltage of the processor may be allowed only when the processor is running below a certain frequency. Such constraints must be closely examined to decide upon the host processor.

Finally, many processors, such as the Texas Instruments OMAP™ series of systems on a chip, have implicit low power or standby modes that should be utilized where possible. Typically in these modes, the processor clocks are stopped (except for a low-frequency, wake-up clock), and as many peripherals as possible are shut down. Readers should note that the "standby" mode used for a Wi-Fi phone (and its host processor) is significantly different from the standby mode of a PC/laptop. In the case of a PC, a transition from the standby to the normal/active mode of operation requires user action. However, a Wi-Fi phone will require the ability to be woken up by external system events as well, as in the case of an incoming call notification. This presents a unique challenge in system design, specifically with respect to integrating these processor standby modes with the operating system. This is discussed in section 9.3.1.4.

As with the WLAN subsystem and the host processor, selection of other peripherals in the system design also needs to take power into consideration. With the widespread popularity of SoC (system on chip) in the telecom industry, this may not be a separate decision and may integrate with the host-processor design choice. However, for some peripherals that exist outside the SoC (e.g., SDRAM, Flash, LCD, etc.), power considerations would still be important.

9.3.1.2 Platform Design

Equipped with the hardware-component selection, the hardware designer must now come up with a board-level design. The underlying design paradigm here is that, from a power-management perspective, it is better to use multiple power rails to source different components instead of using a single power rail in the system. This will allow some power rails to be turned off dynamically when the attached components are not in use, thus saving drainage current. It is therefore absolutely imperative for the hardware designer to understand the interdependence among the different components in the various subcircuits. This knowledge will help him to realize which components can be clubbed together and sourced from a single power rail. For example, the DSP and the AIC (analog interface codec) can probably be sourced from the same power rail since they would most likely be used together. To achieve this power-rail distribution, a power-management chip (such as Texas Instrument's TPS65013), which is capable of supplying multiple power rails while taking a single input source, can be used.

A side effect of multiple power rails in the hardware design is the need for efficient voltage transformations. There are two basic types of voltage converters: low drop-out oscillators (LDOs) and switchers. LDOs are attractive because they are cheap, occupy less real estate, and provide clean (low noise) power translations. However, they can be inefficient, especially in converting from a large voltage to small one. For example, if an LDO is used to convert from the battery voltage (say 4 volts) to a 1.5-volt peripheral in a phone that uses 50 mA of current, the effective drain on the battery from that peripheral will still be the entire 50 mA so that the efficiency is in the 40% range. The potential saving in power is in effect lost in heat. Switches, on the other hand, are more efficient in the voltage translation, providing efficiency in the 80–90% range. However, switchers are more expensive, occupy more real estate (overall) and can cause noise problems because the transformation is not as clean. Therefore, the choice between switches and LDOs is an important design consideration. The platform designer must evaluate the trade-offs carefully on a case-by-case basis.

Another important power consideration for the platform designer is the use of resistor pull-ups and pull-downs in the circuit. Each pull-up/pull-down can be a potential source of drainage current. Though it may not be possible to completely eliminate this drainage current, careful planning with the resistor values (use high-resistance pull-ups on non-speed-critical signals to reduce power consumption) and their placement (scrub circuit for redundant pull-ups/downs; use external pull-ups only when necessary and in sync with internal pull-ups/downs) can minimize this drainage.

9.3.1.3 Software Design

As shown in Figure 9.1, a Wi-Fi handset consists of multiple hardware components. A key observation is that, whereas some components can manage their own power independently without input from the rest of the system, others need to manage their power in coordination with the rest of the system. In our experience, the approach most suitable for a Wi-Fi handset is:

Distribute where possible; centralize where necessary.

This means that if a device is independent enough to manage its own power without input from the rest of the system, then let this device manage its power. Examples of such devices in a typical Wi-Fi handset would include the DSP and the WLAN subsystems. We refer to these devices as independent devices (IDs). It is important to note that IDs are independent from the rest of the components in the handset. Most likely their power-control algorithm and decisions would depend on external events. For example, the DSP may go into a low-power state when no voice processing is required (i.e., no calls are active). Similarly, the WLAN subsystem power algorithm would most likely be tied to network traffic and AP behavior.

On the other hand, there are devices that need to be aware of other components in the system to make decisions regarding power management. Examples of such devices in a typical Wi-Fi handset would be the host processor and the backlight, etc. We refer to these devices as codependent devices (CDs). The existence of CDs in a Wi-Fi handset means that we need a centralized power manager or a power-management module (PMM) to keep track of the system states (SS). A system state (SS) defines what the system is doing at any given instant of time. Depending on the SS, PMM can dynamically configure the CDs in the system to minimize system power consumption.

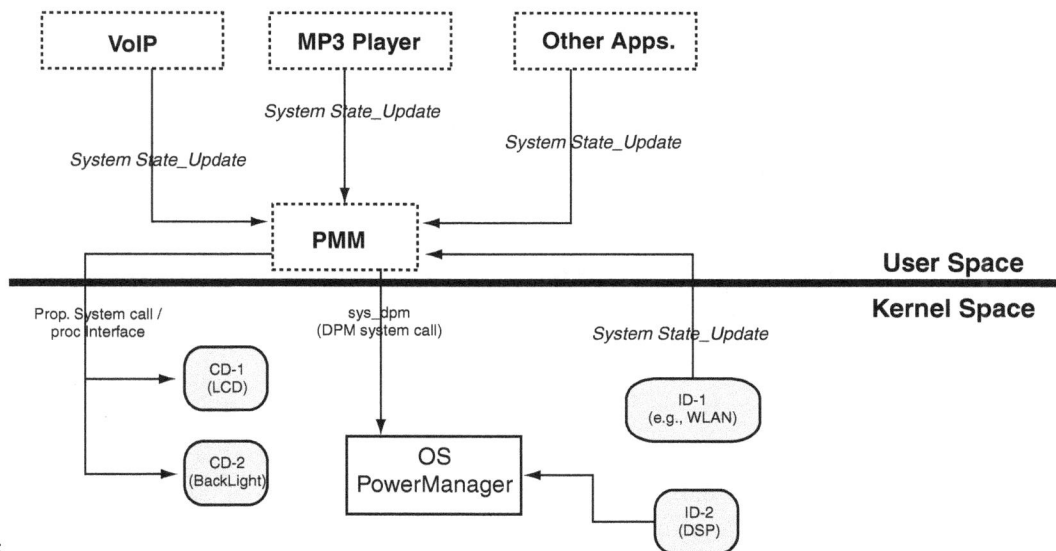

Note:
CD Peripherals are power-managed by PMM.
ID Peripherals manage their own power and inform PMM directly or via OS.

Figure 9.2: Power Management Architecture

Figure 9.2 shows an example of how PMM fits into the software architecture of a Wi-Fi handset. Since the PMM needs to keep track of the SS, it needs to interact with all software components in the system. Specifically, the host processor on which most of the software components execute is a CD (since its load depends on multiple software modules). To control the power of the host processor, PMM needs to know the SS with respect to what each software component is doing at any given instant. For example, the voice application may need to inform PMM about whether or not a voice call is currently active and the MP3 player may need to inform PMM about whether or not a song is being played out.

It is for this reason that software components designed for running on a Wi-Fi handset must be designed keeping in mind the requirement that they need to inform the PMM about their status. It is important to realize that there are significant timing considerations involved here. Consider, for example, what happens if the Wi-Fi handset is in standby mode (with the LCD shut off) and there is an incoming voice call. To display the caller-id, the LCD needs to be turned on. The LCD in turn is a CD whose power is controlled by PMM. The PMM, however, relies on the VoIP application to know about incoming calls. In this scenario, the System-State_Update message from the VoIP module should reach the PMM and the PMM should process this message and turn on the LCD when the phone starts ringing.

Therefore, to make the whole system work in synchronization with the right timings requires that all components issue the appropriate SystemState_Update messages at the right time. Additionally, the priorities of different tasks in the system must be set correctly. Specifically, it may be required that PMM run as the highest priority task in the system so that it can process SystemState_Update messages and wake-up/configure devices as required by the SS.

Another interesting aspect of PMM and software design is the operating system's user-space versus kernel-space design decision. As we discussed, PMM can be implemented as a high-priority user task in the system. However, most PMM implementations would require some part of the implementation to be done in the kernel space as well. Specifically, PMM configuration of CDs for power-management purposes like changing the host-processor clock frequency and LCD power-off and power-on would expectedly require code to execute in the context of the kernel. The interface between the application-level (user-space) PMM and the kernel-level (kernel-space) power manager depends on the operating system in use.

9.3.1.4 The Operating System

A major choice for the system architect is the selection of the Operating System to be used. As always, many parameters must be considered before the selection of an OS. Often these considerations may extend beyond the technical domain and there may be business/marketing reasons for the selection. However, from a power-management perspective, the selection of an operating system is an extremely important decision.

We have worked with the DPM (dynamic power management) framework in Linux and we briefly describe it below to give the reader an idea of how support from the OS can aid in overall power-management architecture.

Dynamic power management (DPM) is a power-management framework that was jointly developed by Montavista and IBM. It is available in 2.6 and above public kernels and from Montavista for 2.4 kernels. It is important to understand what the DPM can and cannot do. DPM is a power-management framework. This means that it provides the basic infrastructure and hooks inside the operating system to be used for achieving power management. This basic infrastructure is platform independent. A platform-dependent part must be implemented for each platform wishing to use DPM. It is this division that allows DPM to be customized for different platforms. There are four principal components of DPM, which are explained below from a Wi-Fi handset perspective.

We start with the host processor where Linux is running. The term "host processor" can be used to refer to a stand-alone microprocessor or (much more likely) an SoC consisting of a microprocessor and a bunch of on-chip peripherals. Each power-aware host processor will provide a number of parameters (various clock frequencies, voltages, etc.) that may be modified at run time to achieve the optimum balance between power consumption and performance. A complete set of values for each of these configurable parameters on the host processor is defined as an "operating point." An operating point therefore specifies all the power-related parameter-values that the host processor on a Wi-Fi handset should be using at any time. The DPM framework is responsible for "moving" the system from one operating point to another depending on the "operating state" of the system.

The "operating state" is the host-processor state that has implications for power management. It conceptually defines what the system is doing at any given time and is closely related to the system state (SS) concept discussed previously. DPM treats the Linux kernel as a state machine that takes the system from one state to another. Even though treating the operating system as a state machine may be highly complex, from DPM's viewpoint it is necessary to distinguish only among those states that may require a change in the operating point of the system. The operating points and the operating states form the basic components in the DPM architecture.

We can now define the concept of an "operating policy." In DPM, the operating policy may be used to define a one-to-one mapping between operating state and an operating point. At the very basic level, it is easy to think of this mapping as the table that describes how fast the processor clocks should be running at any given instant of time in the phone's lifetime. The DPM framework, then, allows a centralized policy manager (like our PMM) to dynamically switch between policies. Note that the intelligence of which policy to use at any given time lies in the user-implemented application-layer policy manager (PMM). The aim of DPM is to provide a framework to move the system from one operating point to another, depending on the operating state, or system state. This flexible design approach, where the OS provides for a power-management framework and relies on the application to drive the power-management state machine, allows DPM to be adapted to various applications and platforms.

Our discussion of DPM would have ended here if we were just concerned with power management on the host processor, since operating points control only the configurable parameters of the host processor. Since the state of off-chip (aka on-board) peripherals/devices has a tremendous influence on the system-wide energy consumption and on the choice of the appropriate operating point, the DPM framework provides for an interface with devices (read off-chip peripheral). Without getting input from these devices, it is almost impossible for DPM to decide what operating point to use. For example, the host processor should not go into its standby mode if the DSP is processing voice packets or if the WLAN subsystem is scanning for connections or if the LCD is being updated due to user interaction. DPM's device interface, therefore, allows devices to assert (and deassert) "constraints." In our example, if the DSP is processing voice packets, it can assert a constraint to the DPM restricting it not to let the host-processor frequency go below a certain threshold. Similar arguments hold true for other device drivers. It is therefore the responsibility of each device driver to assert and deassert constraints to the DPM depending on what the device expects from the host processor.

Given this concept of device constraints, there might be multiple suitable operating points (set of on-chip frequency/voltage values) for each operating state (system state, depending on what the system is doing at any time) depending on the currently asserted constraints (reflects state of devices). For example, on a Wi-Fi phone, in the "task" state, which is used by the entire voice software, the DSP may either need to be run at 50 MHz or at 100 MHz depending on whether one call is active or two calls are active. Therefore, we have two "congruent" operating points for the "task" state, the selection between which would depend on the constraints asserted by DSP. The set of operating points that are congruent with respect to an operating state is known as the operating class. Each operating state is mapped to a corresponding operating class. At any operating state, the DPM parses the corresponding operating class and chooses the first available operating point that satisfies the currently asserted constraints.

9.4 Implementing Power Management

Section 9.3 talked about power management from an architecture and high-level design perspective. In this section, we dig a little bit deeper into power-management implementation.

9.4.1 WLAN Subsystem

The WLAN subsystem is by far the biggest power consumer on a Wi-Fi phone. A significant contribution to the overall system power consumption is due to the radio component of the handset. It is therefore extremely important to minimize the power consumption of the WLAN subsystem. We first look at power-save mechanisms as specified in the 802.11 standard and then go on to look at the standard-independent power-save mechanisms for WLAN.

9.4.1.1 Power Save in Base 802.11 Standards

Section 5.7 describes the power-management techniques described in the base 802.11 standard. Recall from section 5.7 that the 802.11 power save (also known as 'legacy power save') mechanism allowed a sleeping station (a Wi-Fi phone, in our case) to wake up and transmit a packet whenever it was ready. However, the sleeping station had to wait for the AP's beacon to retrieve packets on the receive path. This meant that packets destined from the AP to the station could be delayed for as long as the beacon period. This delay was typically unacceptable for real-time traffic like voice and hence the legacy power-save mechanism was not suitable for VoWLAN. One early workaround was to use a voice codec with large packetization time and an AP with a short beacon period. If the packetization period used for the voice packets and the beacon period were both on the order of 50 ms, it was possible to "hide" this power-save delay by synchronizing the packet transmission with beacons. However, smaller beacon periods meant that the phone had a higher frequency of wake ups from sleep/doze mode when not involved in a phone call. This, in turn, meant higher power consumption when the phone was in standby mode. This caveat made the solution unacceptable for most scenarios.

9.4.1.2 U-APSD Basics

To solve the problem of power management for real-time traffic like voice, 802.11e proposes U-APSD (unscheduled-automatic power save delivery), also known as UPSD (unscheduled power save delivery), which extends the 802.11 doze mode to make it suitable for use in real-time communications.

In U-APSD, the AP need not wait for the TIM (traffic indication map) in the beacon to indicate to the station that it has packets buffered for it; instead it uses a packet received from the station to infer that the station is ready to receive packets. In other words, the voice packet transmission implicitly "triggers" transmission of voice packet(s) waiting at AP. After receiving packets, the station can go back to sleep. As in the base 802.11 standard, the AP uses the "More" bit to indicate whether it has more packets buffered for the STA.

The premise behind the design is that, when a sleeping station wakes up to transmit its packets, it can continue to stay awake in order to receive packets. This approach significantly reduces the delay and jitter in downstream (AP → station) since the station no longer has to wait for the next beacon (which may be as much as a few hundred milliseconds away) to receive its packets. The cost to pay for this reduced delay and jitter is the extra time (and hence extra power consumption) for which the station has to stay awake after transmitting a packet in order to receive packets. However, this time (and hence the associated power consumption) can be reduced to a minimum by ensuring fast turn-around times at the AP.

Note that in U-APSD, the station "need" not wait for the TIM in the beacon to indicate to the station that it has packets buffered for it—however, it "may" wait for this indication in the beacon. This distinction is important to emphasize. The fact that the station need not wait for the TIM in the beacon is useful during a voice call, since voice traffic is periodic and bidirectionally balanced. These two characteristics mean that a station involved in a voice call wakes up every few (typically 20–30) milliseconds to transmit a packet and also expects to receive a packet from the remote endpoint every few (typically 20–30) milliseconds. Hence, when this station wakes up to transmit a packet, it is reasonable to expect that the AP has packets buffered for it from the remote endpoint. Thus, every voice packet transmission from a station can implicitly "trigger" transmission of voice packet(s) buffered at the AP to the station without waiting for the TIM in the beacon.

On the other hand, the station may still use the TIM in the beacon to infer whether there are packets buffered from it is used during call establishment. For an incoming call to the wireless IP phone the U-APSD scheme is not applicable since the station is not actively transmitting any packets and therefore the station (Wi-Fi phone) must continue to rely on traditional power save mechanism.

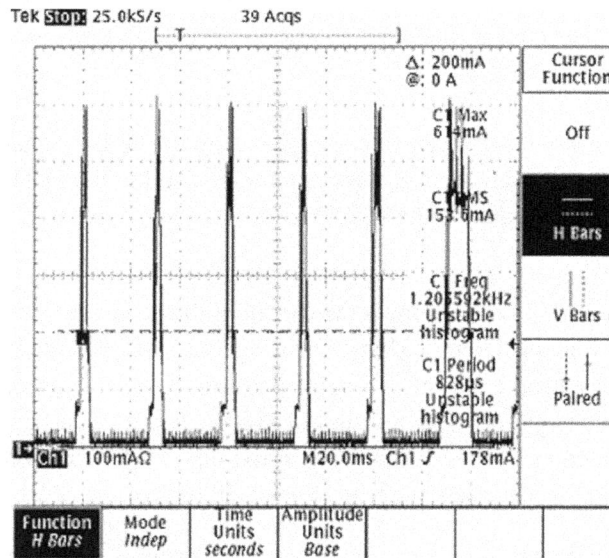

Figure 9.3: Current Consumption of a Wi-Fi Phone During a Voice Call Using U-APSD

Figure 9.3 shows the oscilloscope trace of the power consumption in the system during a voice call using a codec with a packetization period of 20 ms. As can be seen, there is a huge jump in the power consumption (every 20 ms) when the system wakes up to transmit a packet. Figure 9.4 gives a time-expanded view of Figure 9.3, showing what happens during the transmission of a single voice packet. The figure is helpful in understanding the complete process. The system power levels go through various highs and lows as the system:

a. wakes up and prepares (processes packets) for transmission

b. transmits a voice packet

c. waits a short while for a packet from the AP

d. receives the voice packet

e. does some post-processing

f. goes back to low-power mode until the next voice packet is ready to be transmitted.

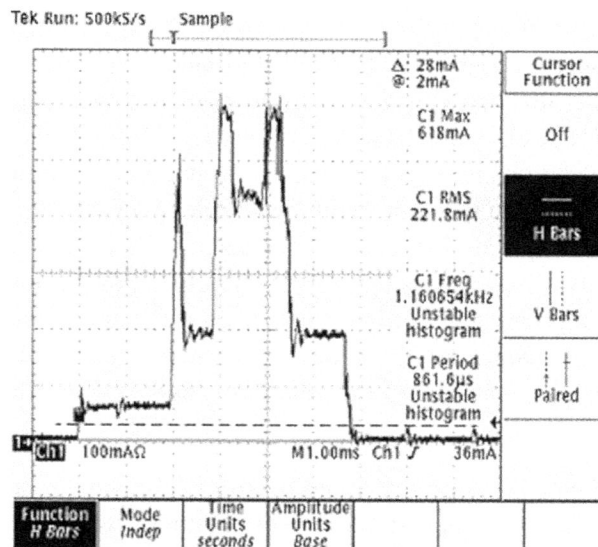

Figure 9.4: Current Consumption of a Wi-Fi Phone During the Transmission of a Single Voice Packet

9.4.1.3 U-APSD Warnings!

Realize that U-APSD reduces the problem of large variable delays (when using power save) in the voice stream by introducing a small jitter into voice flow. This is somewhat like the concept of quantization. By triggering the packet retrieval from the AP every p milliseconds (where p is the packetization period), we are ensuring that no packet gets delayed by more than p milliseconds. However, this also means that packets would have to wait at the AP for an average of $p/2$ milliseconds. Note that this jitter would not have occurred if the station had been in active (always power-on) mode all the time. However, most Wi-Fi phones would be happy to trade off a jitter of $p/2$ milliseconds for the substantial gain in power save that U-APSD brings.

Next, note that in U-APSD the retrieval of the packets from the AP is triggered by the transmission of the packets from the station. This approach works because of the bidirectional nature of voice traffic. However, this assumption of bidirectionality gets broken in certain circumstances. This may happen, for example, if VAD[3] (voice activity detection, or silence suppression) or call hold or local mute is used with U-APSD during a voice call. In these circumstances, the downstream packets buffered at the AP may actually be delayed since the station is not transmitting anything. Obviously, this is unacceptable. Therefore, the Wi-Fi phone must ensure that, even during periods of silence, some (Null) data packets are transmitted from the station to the AP so as to ensure the triggering of voice packets in the downstream. This solves the problem of U-APSD and VAD coexistence, though at the cost of some extra bandwidth consumption in the station–AP link.

Another interesting scenario is the use of U-APSD in roaming scenarios. To reduce roaming times (which is important for Wi-Fi phones), stations often use background[4] scanning techniques to keep track of "currently available APs to roam to." Note that scanning means that the radio has to be powered on and kept powered on for hundreds of milliseconds. However, U-APSD works by powering on the radio right at the time of a packet transmission and then powering it off as soon as a packet from the AP is received. This whole process is usually less than 10 milliseconds, which means that there is no time to do scanning. Here, one solution is to split one big scan into multiple small scans by scanning one channel at a time. Furthermore, these single channel scans can be synchronized with voice packet transmission to ensure that the radio is powered on for the minimum possible time while still maintaining small roaming times.

A related issue is the choice between active and passive background scans when using U-APSD. Even though passive scan may seem to be an attractive option in terms of power savings (since it does not involve any transmission), the close timings involved in U-APSD make active scans a better option. Since active scans involve probing the AP for a response rather than waiting passively for a beacon, they can potentially finish faster than passive scans, thus minimizing the delay and jitter caused by background scans during voice calls. It turns out that this strategy is better even from a strictly power-save perspective. Active scans may consume more power per se due to the transmissions involved, but since they tend to finish faster, the overall power consumption is lower.

---

[4] To save bandwidth, VAD is used to stop (or reduce the rate of) packet transmission during periods of silence.

[5] Recall that, in the context of a Wi-Fi phone, "background scanning" refers to the act of scanning done while a voice call is ongoing.

9.4.1.4 U-APSD and Access Categories

The previous sections have provided U-APSD in some detail. To recap, the basic design of U-APSD is simple to describe:

For the station:

   a.   Station informs the AP that it is going to sleep by setting the Power Save bit in the 802.11 header just as in 802.11 legacy power-save mode.

   b.   Station wakes up from its sleep mode to transmit a packet whenever it needs to. The transmission of this packet marks the beginning of an unscheduled SP (service period).

   c.   Instead of immediately going back to sleep, the station continues to stays awake till it receives a packet from the AP that has its EOSP (end of service period) bit set.

For the AP:

   d.   AP buffers packets for sleeping stations.

   e.   When the AP receives a packet from a sleeping station, it transmits N buffered packets to this station. All, except the Nth packet, transmitted by the AP to the station must have their EOSP bit clear, whereas the Nth packet must have the EOSP bit set. This allows the station to determine when it can go back to sleep.

   f.   The minimum allowed value of N is 1, which means that even if the AP has no packets buffered for the station, it must send a Null data frame with the EOSP bit set so that the station may go back to sleep.

   g.   The maximum allowed value of N is configurable by the station; i.e., each station may decide whether it want to receive 2, 4, 6 or "all" the packets that are buffered at the AP during an SP.

Unfortunately, the actual U-APSD protocol is a little more complicated than this. The complication arises from the existence of multiple ACs introduced for achieving QoS. To appreciate the problem, consider a scenario where a station is involved in a voice call and is also running a web-browser to download a web-page from the Internet. In this scenario, the station is using both its voice AC and its best-effort AC.

Now, if this station wants to use U-APSD, it will be waking up periodically to transmit its voice packets. On receiving the voice packets, the AP should transmit the packets it has buffered for this station. However, we have a choice here: should the AP use this "trigger" packet to transmit only those packets that are in its voice AC or should it use the trigger packet to transmit all packets that it has buffered for this station irrespective of the AC? This is not an academic dilemma. It is very likely that the AP has packets buffered for this station both in the voice AC (from the remote phone) and in the best-effort AC (from a web-server). U-APSD solves this problem by leaving this choice up to the station.

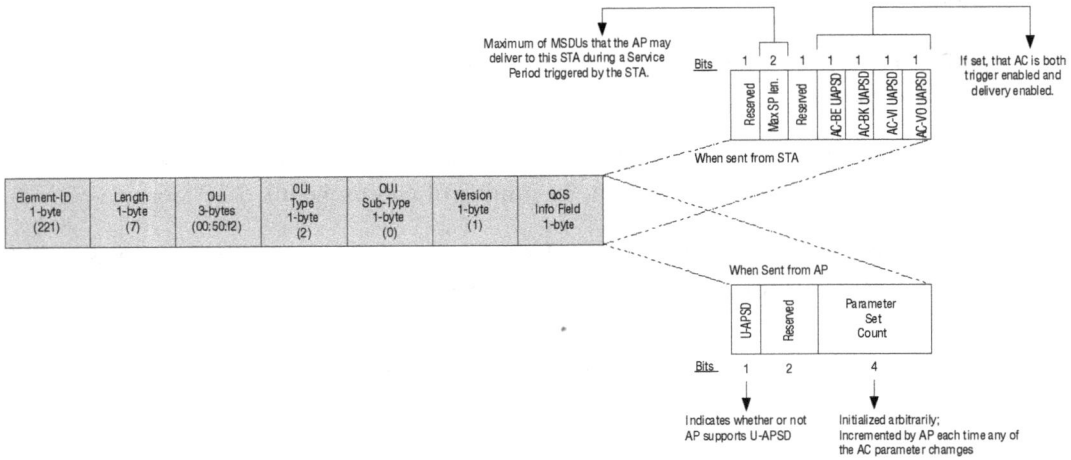

Figure 9.5: Element ID 221

The station can request the AP to make one or more of its ACs "delivery enabled." This can be done using element ID 221 (see Figure 9.5), which can be part of the Association frames or the TSPEC frames. Now, when the AP receives a trigger packet from a station, it will transmit buffered packet(s) to that station only from those ACs that are delivery enabled. Packets buffered in other ACs will continue to be delivered using legacy power save—i.e., by setting the appropriate bit in the TIM in the beacon and then waiting for a PS-POLL from the station.

There is another aspect of U-APSD and multiple ACs that makes the protocol even more complicated. Continuing with our example, where the station is using both its voice AC and its best-effort AC simultaneously, suppose that this station has requested the AP to delivery enable all its ACs. This means that, on receiving a trigger packet, the AP will transmit buffered packet(s) to that station from all ACs. However, what qualifies as the trigger packet? Is a packet transmitted from the voice AC the trigger packet, or is a packet transmitted from the best-effort AC the trigger packet, or can any one of them be considered a trigger packet? Again, U-APSD solves this problem by leaving this choice up to the station. The station can request the AP to make one or more of its ACs "trigger enabled." This can be done using element ID 221 (see Figure 9.3), which can be part of the Association frames or the TSPEC frames. Only a packet from a trigger-enabled AC can be treated as a trigger for starting downstream data transmission.

To summarize, the design of U-APSD is made complicated by the existence of multiple ACs. Since each AC can be independently trigger- and/or delivery-enabled, multiple configurations can exist, which lead to different behaviors. Network administrators must closely examine the needs of their network and then choose the appropriate configuration of U-APSD.

9.4.1.5 WMM-SA Power Save

As we discussed briefly in Chapter 6, there is a second QOS-aware scheduling mechanism for 802.11, referred to as the hybrid coordination function (HCF), or WMM-SA (scheduled access). WMM-SA also defines a power-save mode suitable for voice, called scheduled automatic power save delivery (S-APSD). S-APSD is set up via TSPECs at the same time that the SA traffic flow is defined. It is in fact a simpler scheme because of the scheduling aspect of WMM-SA. Essentially, the AP returns, in its TSPEC response, a time offset from the beacon where the requested service period is to start. A Wi-Fi phone can sleep until this offset, wake up to send and receive data, and then sleep until the next service period (e.g., 20 ms later).

9.4.1.6 WLAN Power Save Beyond Standards

Until now, this section has discussed standards-based power-save mechanisms. These mechanisms discuss power management from a system-architecture perspective. In other words, the standards involve themselves with power management coordination between stations and Aps: for example, when does the station go to sleep, for how long and how does it wake up? However, these standards do not specify how the station should implement doze/sleep mode. This decision is left to the station designers. This means that each station is free to choose which components to shut off when going into doze mode. It is expected that, at the very minimum, power-efficient stations would turn off their radio components when in doze mode. Further power savings during doze mode can be achieved by turning off other components in the station.

In a typical implementation, WLAN software implementation is distributed between host processor and external chip sets (e.g., MAC and Radio). Therefore, power management of the WLAN subsystem too is handled by the firmware on the Wi-Fi chipset and the WLAN driver on the host processor. Since the power management of the WLAN subsystem must be closely synchronized with the AP timing, as per the standards, low-level aspects of WLAN power-management/control must be localized to the WLAN subsystem itself. From a PMM perspective (see section 9.3.1.3), therefore, the WLAN subsystem is an independent device that manages its own power consumption but keeps the PMM updated about its status, so that PMM can keep track of the SS with respect to radio activity and wireless connectivity to the AP.

9.4.2 LCD and Backlight

The LCD and the backlight are usually the next biggest power consumers after the WLAN subsystem in a Wi-Fi phone. LCD power-control mechanisms are pretty well established in the cell-phone industry and can be reused here. Most algorithms are based on a user-activity timer to turn these components off when possible. This involves tracking when the user is using the LCD based either on keypad interaction or explicit input from applications like a video player, etc. Based on this information, the LCD and backlight can be controlled.

If constant LCD display is required (e.g., for displaying time even when the phone is not in use), it is useful to consider an LCD with a "Still" mode capability, which allows a static image to be displayed without constant refresh from the SDRAM-based frame buffer. This reduces the power consumption.

9.4.3 Host Processor

The host processor is an interesting candidate to consider for power management. The high-level software architecture issues discussed in section 9.3.1.3 and the OS perspective of power management discussed in section 9.3.1.4 form the backbone of power management on the host processor. To summarize, software (device drivers, applications, etc.) on the host processor must coordinate among themselves (if possible, via the OS) to optimize power management and performance.

It is worth emphasizing, however, that most commercial software support for standby modes (of systems/host processors) is mostly slanted towards PC/laptop-type standby. This is likely not appropriate for a Wi-Fi phone for the following reasons:

1. A Wi-Fi phone must be able to get in and out of the standby mode(s) "quickly."
2. Transitions to and from standby mode(s) must not lose system state, e.g., we do not want to reload system software from flash after every standby transition.
3. System must have a real-time clock active even during standby.
4. Besides user interaction, the system should exit standby mode on:
   a. network activity (incoming).
   b. application-specified timer from real-time clock; this also means that architects should make sure that application timers don't cause unnecessary wakeups.
   c. other. (requirement specific).

9.4.4 DSP and Analog Codec

The DSP and analog codec are combined together from a power-management perspective, since they are most likely to be used together. The first step for these components is to shut down DSP (and analog codec) when they are not required—that is, when there is no voice call active. The system designer should be aware that this technique may come at the cost of added delay at the time of call setup, since powering on the DSP may require the whole DSP code to be downloaded before processing can start. This would add to the delay for call setup. Some DSPs may provide a state where the DSP can be put in standby mode with minimum power consumption, with the advantage that the DSP code is maintained and does not need to be redownloaded when transitioning from the standby state to active state.

When the DSP is powered on for use during a call, the software can adjust DSP clocks based on processing requirements. For example, the DSP clock-frequency requirements would probably depend on the codec to be used, the echo-cancellation tail length required, etc. Running the DSP clock only at the least possible clock frequency would lead to power savings. Another power-saving technique is to place the DSP idle loop in the fastest memory to optimize its idle loop. This is beneficial both during a call and when no calls are active.

Similarly, many analog codec ICs actually come with multiple codecs. If only one voice stream is active, the other analog codec can be powered off, thus leading to power savings.

9.4.5 Memory

"The memory sub-system in a Wi-Fi phone can be broadly divided into two categories: External & Internal memory. The former refers to the memory external to the host processor (SoC) and the latter refers to memory internal to the host processor (SoC). External memory typically consists of a Flash (contents retained at power off) and a SDRAM (contents lost at power off). Internal memory on the other hand consists of internal SRAMs and caches.

Besides selecting low power memory components, from a power management perspective, the following guidelines should be kept in mind to optimize the power consumption of the memory sub-system.

- SDRAM can be put in self-refresh mode when not in use for example, this is typically do-able when the phone is in standby mode.

- Frequently accessed (Active) code and data should be moved into the internal chip memory to minimize accesses to external memory—which typically consume more power than internal accesses.

- As an extension of the above, the idle-loop of the processor should be in the internal memory.

- Caches should be used effectively to reduce external memory accesses.

- Effect of page swapping should be evaluated and certain pages can be locked into memory, if optimal.

9.4.6 Other Peripherals

We have covered the power-management techniques of some of the most important peripherals in the preceding sections. Undoubtedly there are numerous other power consumers in a Wi-Fi phone. However, a complete discussion of all peripherals is beyond the scope of this book. We hope that the preceding sections have provided the reader with a broad overview of the philosophy and techniques of power management, which can be extrapolated for other components. As mentioned earlier, the general philosophy is simple:

    a.   When not in use, shut it off.

    b.   To use, turn on only as much as necessary.

9.5 An Operational Perspective

Thus far, we have talked about power management from a design and implementation perspective. It is now time to take a step back and think about the user effect of power management. From, an end user's viewpoint, power management is evaluated in terms of "talk time" and "standby time." Though commonly used, these terms are rarely well defined. We will use these terms according to the following definitions.

Talk Time: Refers to the time duration for which the user can use the phone to "talk." This can be calculated as the single longest duration phone call possible. The talk-time numbers as quoted by various vendors in the industry are calculated under ideal talk time conditions to get the best possible number for advertising. These ideal conditions include:

    a.   There is no other application running except the voice application.

    b.   The user is at a constant distance from the AP and is not roaming.

    c.   The user is not using the key-pad interface while she is talking.

    d.   Optimum call parameters (codec, echo tails, etc.) are being used.

Standby Time: Refers to the time duration for which the user can "use" the phone without recharging the battery. This is usually calculated as the longest time the phone is "alive" (ready to make/receive calls). Again, the standby numbers as quoted by various vendors in the industry are calculated under ideal standby-time conditions, to get the best possible number for advertising. These ideal conditions include:

    e.   No voice calls are being made while in the standby mode.

    f.   There is no other application running except the voice application.

    g.   The user is at a constant distance from the AP and is not roaming.

    h.   The user is not using the key-pad interface.

With these terms defined, the aim of power management is to be able to maximize both the talk time and the standby time. The following sections discuss how to achieve this.

9.5.1 Maximizing Talk Time

As explained in the previous section, talk time is usually calculated as the single longest duration phone call possible. Under these ideal conditions, the state of various devices on the Wi-Fi phone in this mode is discussed below.

Radio/WLAN subsystem: The power management of this device/subsystem is governed by the U-APSD standards as discussed in section 9.4. Basically, the fixed size and periodic nature of the RTP packet stream allows the WLAN subsystem to be switched off for the time duration between voice packets and therefore makes U-APSD well suited for VoIP.

Host Processor: Under these conditions, the host processor's work is usually limited to transferring packets from the DSP to the network. Hence, the host processor's frequency may be lowered significantly to save power. Furthermore, depending on the capability of the host processor, it may also be possible to lower its operating voltage.

DSP and Analog Codec: These components must be powered on since they have to do signal-processing functions like coding-decoding functions, echo cancellation, etc. However, power-save mechanisms mentioned in section 9.4.4 can still be used to optimize their power consumption.

LCD and Backlight: Can be switched off since typically user interaction with the phone is minimal during a voice call and, in any case, a key press can be used to turn the LCD and the backlight back on.

Peripherals: The on-chip and off-chip (on-board) peripherals have to be handled on a per case basis, but it is expected that a large majority of them could be powered off during a voice call.

Drainage Current: The drainage current would continue to exist at talk time. Good platform design (see section 9.3.1.2) can, however, minimize it.

It is important to note that since many of the active components in the system synchronize their transitions between low-power and high-power modes to voice packet transmission, the packetization period of the codec used during the voice call is an important factor affecting talk time. Larger packetization period means that all processing components (WLAN sub-system, host processor, DSP, etc.) have to go to high-power modes less frequently from their low-power modes. This means lower power consumption from a system perspective. The obvious trade-off is that larger packetization periods lead to higher delay and jitter, which may negatively affect voice quality.

It has also been argued that codecs with high compression ratios are better for power save since they manage to pack more information into a smaller number of bits, which means that overall the WLAN subsystem has to wake up for a smaller duration to receive and transmit the voice stream. This, in turn, leads to lower power consumption. Even though this argument has some weight, in practice the header sizes and fixed overhead tend to dominate the packet size so that power improvement from a low-bit-rate codec is not that great. Furthermore, high-compression codecs may also need to run the DSP at higher frequencies (low-bit-rate codecs require more MIPS), which may mean more power consumption for the DSP.

9.5.2 Maximizing Standby Time

Section 9.5 defines standby time and the ideal conditions under which it is usually measured. Under these ideal conditions, the state of various devices on the Wi-Fi phone in this mode is discussed below.

Radio/WLAN subsystem: It should be noted that power consumption due to the WLAN subsystem waking up may affect more than just the WLAN chip set. Specifically, the frequency of wake-up will affect Host Sleep Mode effectiveness. Hence, this is a significant area of impact.

When the phone is in standby mode, the WLAN subsystem can mostly be in doze/sleep/off mode but must wake up periodically to process beacons and broadcasts. The challenge here is to minimize the wake-up frequency and the wake-up duration. The wake-up frequency is a factor of the AP beacon period (a network setting), the AP DTIM period (a network setting), and the amount of broadcast traffic in the BSS (a network traffic characteristic). Longer AP beacon and DTIM periods increase the battery life but at the cost of added delay in the network. This is a trade-off that network administrators should make. Some optimizations are possible here. Longer DTIMs, for example, will have adverse side effects in terms of ARP delays. However, proxy ARP can help here. By having the AP respond to ARP requests for stations associated with the AP, the station need not wake up to process every ARP request. This means lower wake-up frequency for stations like Wi-Fi phones, and hence longer standby times.

Host Processor: It should be in sleep/standby-mode but is periodically woken up by external events (incoming call, beacon processing, key press, etc.) or by timer interrupt from an RTC (real-time clock).

DSP and Analog Codec: Shut off.

LCD and Backlight: Shut off.

Peripherals: Shut off.

Drainage Current: The drainage current is minimized by turning off most power rails. The only power rail(s) that are expected to remain powered on are the ones feeding the RTC.

9.6 Summary

We have discussed power management for a Wi-Fi phone from a system perspective in this chapter. The aim has been to give the reader a broad overview of the topic. Given the scope and depth of this topic, it is very difficult to cover it in great detail. Each design will have its own issues and trade-offs to be made. We hope that this chapter serves as good guidance for designers of those products.

Voice over Wi-Fi and Other Wireless Technologies

10.1 Introduction

So far we have been discussing voice over Wi-Fi in the context of pure VoIP technology, running over 802.11a/b/g-based networks. In this chapter we will look at voice over Wi-Fi in conjunction with other wireless technologies, including proposed 802.11 extensions, cellular, WiMax (and other wireless broadband technologies), Bluetooth, and conventional cordless phone systems.

The field of telecommunications is in flux. Probably the most often-used word today in telecommunications is convergence. VoIP has been a big step towards the convergence of the voice (PSTN) and data (IP).

Till very recently, the C-word was limited to the discussion of wired networks and [cellular] wireless networks, and the discussions were dominated exclusively by voice-oriented wireless networks where IP had not make any significant inroads. With the emergence of Wi-Fi, IP has now gone wireless. Furthermore, the emergence of VoWi-Fi means that industry pundits are talking about the convergence of voice (GSM, 3G) and data (Wi-Fi, WiMax) in the wireless domain too. But what does all this mean? What will the wired and wireless networks of tomorrow look like? Which technology will "win"? We look at these issues in this chapter.

10.2 Ongoing 802.11 Standard Work

One characteristic of the 802.11 standard is the ever-present enhancement process. Some of these enhancements are just now coming into the market, but there are others still under specification. This section summarizes the ongoing work and the possible impact on a voice application.

Table 10.1 outlines the various 802.11x projects as a reference.

Table 10.1: Summary of 802.11 Projects

802.11 Project	Description	Status (as of January 2006)	Impact to Voice
.a	Up to 54 Mbps in the 5-GHz range. Introduces the OFDM modulation technique.	Ratified; products in the market.	As discussed earlier, 802.11a is important because it adds more available channels and thus increases 802.11 voice-carrying capacity. Some impact to scanning and roaming algorithms (see Chapter 8).
.b	DSSS transmission rates in the 2-GHZ ISM band. This is the base standard.	Ratified.	
.c	Bridging operation. Not a standalone standard; work merged in to 802.1d.		N/A
.d	Global harmonization	Ratified.	Discussed briefly in Chapter 4.
.e	Quality of Service	Ratified (Wi-Fi variants WMM and WMM-SA).	As discussed in Chapters 6 and 9, WMM provides two important features for the voice application: • Prioritization of voice over other traffic • Power save method that is useful for voice traffic
.f	Inter-AP protocols	Ratified, but never really accepted in industry.	N/A
.g	Enhancement of transmission rates to 54 Mbps.	Ratified.	Higher capacity means more voice calls per AP.
.h	Regulatory enhancements and spectrum management for .a.	Ratified.	Some impact to the scanning algorithm as discussed in Chapter 8.
.i	Security enhancements.	Ratified. Wi-Fi variants: WPA and WPA2.	See Chapter 7.
.j	Japanese Regulatory—defines frequency requirements for 4.9-GHz and 5-GHz deployment in Japan.	Ratified.	No real impact.
.k	Measurement enhancements.	Under development.	Chapter 8 touched on some of the proposed .k features and their impact on roaming and scanning.
.l			No project.

Table 10.1: Summary of 802.11 Projects (continued)

802.11 Project	Description	Status (as of January 2006)	Impact to Voice
.m	Editorial cleanup of specifications.	Ongoing.	N/A
.n	Enhancement of transmission rates to wide bands throughputs.	Under development.	See text below.
.o			Not a project.
.p	Wireless access in a vehicular environment.	Under development.	See text below.
.q			Not a project.
.r	Enhanced roaming.	Under development.	Discussed in Chapter 8.
.s	Mesh networking.	Under development.	See text below.
.t	Wireless performance prediction and test.	Under development.	See text below.
.u	Internetworking with external networks.	Under development.	See text below.
.v	Wireless network management.	Under development.	This project is looking at ways for APs to actively control station radio settings; for example, through network management protocols such as SNMP. It is closely tied with the 802.11k project.
.w	Security for management frames.	Under development.	This project is looking at extending 802.11i to protect management frames.
.x			Not a project (.x is used to refer to the entire 802.11 protocol umbrella).
.y	802.11 in the 3.65–3.7-GHz wavelengths (opened up in the US in July 2005).	Under development.	New spectrum means, of course, more potential capacity for voice traffic. There will also be a roaming/scanning impact as per mixed 802.11a/b/g solution today. 802.11y will also include a standard mechanism to avoid spectrum interference with other users of the spectrum. This, in theory, will simplify opening up additional frequency bands in the future.
.z			No project.

10.2.1 802.11n

The 802.11n project is working on techniques to (a) increase the user throughput of 802.11 to over 100 Mbps and (b) increase the range of communication. This will be accomplished through the combined use of several technologies. One of the main technologies is the use of multiple input, multiple output (MIMO) antennas for send and receive. With MIMO, the transmitter and receiver can take advantage of the inherent reflections (multipath) that occur during radio transmission instead of being negatively impacted. The sender and receiver essentially generate/receive multiple data streams (up to four, two typically) that are spatially separated. Using sophisticated mathematics, the receiver is able to recover the data streams. The advantage of this approach is that the data rate can be increased and the range enhanced because the receiver can recover the signal better under noisy conditions.

802.11n also includes an improved OFDM modulation technique that can handle wider bandwidth and coding rate.

Another technology is the use of wider channels: instead of 20-MHz channels for 802.11 a/b/g, 802.11n can make use of 40-MHz channels. This is an optional feature.

Packet aggregation and bursting techniques along with a more sophisticated block-acknowledgment scheme and a smaller interframe spacing (RIFS) are also used to minimize the overhead per data packet transmitted. Aggregation is especially important in mixed mode, where 802.11n devices must coexist with legacy 802.11b/g devices. 802.11n also defines an optional mode (Greenfield) where legacy support is not required. It is in this mode where the maximum data rates will be achievable (in mixed mode, for example, an 802.11n device will at a minimum need to bracket each high throughput with a legacy transmission such as RTS/CTS to clear the medium).

The standard will also include enhanced power-management techniques in addition to the U-APSD method we discussed in the previous chapter. One important, mandatory technique, referred to as MIMO power save, allows the more power-intensive MIMO mode to be disabled for transmissions that do not require the high throughput—i.e., voice. Without this, MIMO operations, at least for initial chip sets, will be more costly in terms of battery life than mature 802.11b/g solutions.

A second, optional technique, power save multi poll (PSMP), uses a microscheduling technique to manage the channel efficiently. PSMP comes in two flavors, scheduled and unscheduled. Scheduled PSPM works in conjunction with scheduled access 802.11e QoS (refer to Chapter 6). At a high level, an 802.11n Wi-Fi phone will create a scheduled PSMP traffic stream, via a TSPEC, at the start of a call. A PSMP frame will be sent by the AP according to the service period (i.e., packetization period used in the call), and will be synchronized to the scheduled power save (S-APSD) interval that is being used. PSMP frames are management frames that are sent to a broadcast destination. Each PSMP frame contains multiple station-information fields, where a particular field gives a station ID (AID) and time offsets when the station is allowed to send and receive and for how long. In effect, these frames are minischedules for the medium (up to 8 ms at a time). The Wi-Fi phone will wake up to receive the PSMP frame and then know when to send/receive its voice packets.

With unscheduled PSMP, the 802.11n Wi-Fi phone will use U-APSD (as discussed in the previous chapter). The U-APSD configuration may or may not have been set up via a TSPEC. In either case, the receipt of a trigger frame will cause the AP to issue a PSMP frame to schedule the transmission of queued frames from its delivery-enabled access category queues. The PSMP will also schedule when the Wi-Fi phone will issue acknowledgments.

To further improve both performance and power saving during PSMP, a new acknowledgment scheme known as Multiple TID Block ACK (MTBA) is used. This ACK scheme allows ACKs (a) to be delayed, and (b) combined together so that all packets received in the PSMP period can be acknowledged with one acknowledgment.

The impact to the voice-over-Wi-Fi application of 802.11n is obvious. Greater data throughput means more calls can be carried per access point and voice/data can be more easily mixed on the same network. One issue with 802.11n handsets will be reduced battery life due to the higher power requirements for 802.11n functions. The MIMO power-save mode will hopefully alleviate some of this and make the power consumption for 802.11n VoWi-Fi devices comparable to today's 802.11b/g solutions.

10.2.2 802.11p

This IEEE project is looking at the changes necessary for 802.11 to operate in the 5.9-GHz licensed intelligent transportation system band (ITS). The focus is on vehicular applications such as toll collection, safety, and commerce. One aspect of the project that is relevant to voice is the requirement for fast (i.e., vehicle speed) handoffs. The approaches being defined in this project may impact the fast roaming work done in 802.11r and vice versa.

10.2.3 802.11s

The purpose of this project is to standardize the protocols for an 802.11-based wireless distribution system (WDS). The proposed architecture for this system is a meshed network that is wirelessly connected using 802.11 protocols. Nodes on the network will dynamically discover neighbors, and enhanced routing protocols will be present to facilitate efficient packet routing. 802.11s compliant networks will be self-organizing.

A standardized wireless distribution system for 802.11 will increase the coverage capability of 802.11 networks, thus making the applicability of 802.11 voice services even more compelling. Instead of being restricted to disjoint islands of Wi-Fi hot spots, one can envision a 802.11s-based mesh network covering large (cell-phone scale) spaces. However, the 802.11s protocols will need to ensure that the quality of service and security concerns for voice that we have discussed earlier are addressed. Furthermore, the per-[voice] packet latency introduced by a mesh topology will have an impact on the overall quality of service that is achievable.

10.2.4 802.11t

The 802.11t project is concerned with performance measurement. This project will result in recommendations (not standards) in the areas of measurement and test techniques and metrics to be tracked. This project and a related Wi-Fi voice test project are interesting in that they recognize that voice-over-802.11 performance test requirements are very different from traditional data performance. Data-performance testing is mostly concerned with throughput, and the metrics of interest are maximum bits per second that can be transmitted or received under various conditions (such as data packet size).

Voice performance, however, is more concerned with the packet loss, latency and jitter that are present under overall BSSS loading. Artificial test and measurement is a difficult problem because to really assess the impact of all three performance areas on a voice call will require incorporating the techniques used in VoIP to mitigate network impacts. For example, as we discussed in Chapter 3, VoIP terminals will use a jitter buffer to handle voice packet interarrival variance, and will utilize packet-loss concealment techniques to recover from lost packets. These will need to be factored into the measurement process.

Power management and battery life are important 802.11 phone metrics. As we discussed in Chapter 9, these are impacted heavily by access-point performance such as beacon interval stability, and ps-poll/null-frame response times. Test methodologies need to be developed for these and other power-related metrics.

A final area of voice-unique testing is in the area of roaming. A standardized test methodology to measuring roaming and handoff performance is a highly desirable product of this project.

10.2.5 802.11u

The 802.11u project is aimed at adding features to 802.11 that facilitate interoperation in multiple 802.11 network environments. These features include network enrollment, network selection and service advertisement. In general, the results of this project should facilitate the Type C roaming that we discussed in Chapter 8. 802.11u will have some overlap with another IEEE project, 802.21. We will discuss this further below.

One important feature being addressed in the 802.11u project is the handling of emergency calling. One proposal being considered is to use a spare bit in the existing TSPEC element to indicate that the requested resources are to be used for an emergency call. A phone could then issue an ADDTS message to the AP with this "emergency service" bit set when a "911" call was placed.

10.3 Wi-Fi and Cellular Networks

Wi-Fi/802.11 is, at its basic level, a radio technology. This section will examine how voice over Wi-Fi will interact with the current "reigning" champion of voice/radio technology: today's cellular networks.

Wi-Fi and cellular are for the most part complementary technologies. They individually provide solutions to a subset of the wireless space. In particular, Wi-Fi can be used to provide coverage in areas where cellular is less effective: indoors, hospitals, airports, and urban canyons.

The inclusion of a voice-over-Wi-Fi capability into a cell-phone handset can be viewed as the ultimate goal for voice over Wi-Fi. There are several reasons for this goal. An obvious one is economy of scale. Economically, a voice-over-Wi-Fi implementation (chip set/software solution) will reap immense benefits from deployment in the 500-million plus cellular-handset market. Such volumes allow for the research and development necessary to create new classes of system on a chip specifically tailored to meet conventional cellular and Wi-Fi requirements. We should expect chip sets in future that include Wi-Fi radios, MAC/Baseband integrated with cellular modems and processors. With this integration comes a reduction in cost that will surely spill over into pure voice-over-Wi-Fi space as well.

Secondly, like its parent technology VoIP, voice over Wi-Fi will piggyback on the ongoing improvements to the data networks that are being upgraded to provide enhanced data services to the basic cell phone. Examples of this include the 3G data networks that have been coming online in the past five years.

A third factor is the usefulness of Wi-Fi technology to augment areas where cellular technology is lacking. Specifically, areas where cellular coverage is problematic (in doors, airports, hospitals, urban canyons) can be handled by overlapping Wi-Fi networks. The introduction of Wi-Fi-based mesh networks, driven by the maturation of 802.11s, will contribute to this trend. For example, vendor studies have shown that a city-wide, Wi-Fi mesh network can be a more cost-effective approach to providing wireless coverage than deploying 3G (1xEV-DO). A Wi-Fi mesh network could be, for example, situated in street lamp posts (which can be rented cheaply from the city government) as opposed to a cell tower that would require space from an office building or home.

A fourth factor is bandwidth. The cellular network providers would love to have the ability to move customers off their precious spectrum wherever possible. For example, when in your broadband-enabled home, why not let the cell phone make calls over the IP-based broadband network, via your in-home Wi-Fi infrastructure? A side effect of this is that cellular providers, riding on top of broadband access, now have a means to get customer phone minutes when he is at home. By providing an integrated access point and VoIP gateway equipment that allows the customer's conventional home telephony (i.e., POTs phones) to place VoIP calls back to the cellular base network, just like with his dual-mode cell phone, the cellular providers can cut into the traditional home voice-service monopoly held by the local telephone companies. This is one of the key drivers of fixed/mobile convergence, with the goal being to get customers to sign up to an all-inclusive home and mobile phone service from the cellular providers.

A final factor is the cellular world trend to move to using an SIP-based call-signaling protocol known as IP Multimedia Subsystems, or IMS. We will discuss IMS further below. The use of SIP to control cell-phone call signaling as well as voice-over-Wi-Fi signaling makes it easy (relatively speaking) to architect a unified phone with seamless handoffs between the cell world and the Wi-Fi world.

10.3.1 Dual-Mode Issues

There are several issues, however, to overcome before the dual-mode, cellular and Wi-Fi phones become a reality. These include:

- Handoffs between the two networks. This is especially a problem if the existing cellular signaling mechanisms are used while on the cellular network and VoIP signaling is used when in Wi-Fi mode. One approach is to simply not allow switchover while a call is in progress. Another is to use the same signaling protocol for both networks. We will look at this case further below.

- Billing. Two networks means two billing systems, assuming that the Wi-Fi portion is not free.

- Phone integration. Integration of a dual-mode phone is a nontrivial exercise. One area of difficulty is the reuse of key hardware and software components. For example, today's cell phone utilizes highly optimized systems on a chip, including possibly accelerator hardware for audio codecs, echo cancellation and other number-crunching algorithms that are executed on the voice samples (refer to Chapter 3). These may be highly integrated with the cellular voice processing, so reusing them when in voice-over-Wi-Fi mode may be difficult.

- Power management. Today's cell phones achieve their battery life levels through a combination of power-efficient hardware and power-aware protocols. For example, the cell-phone voice-sample processing subsystem (codec, echo canceller, etc.) is closely tied to the cellular network "timeslot" so that the entire phone can wake up out of a low-power state only when needed to process, send and receive samples.

A Wi-Fi phone, in contrast, does not have such a close coupling between the voice-processing subsystem and the actual Wi-Fi network. Also, a dual-mode phone will require Wi-Fi channel scanning (and its associated power requirements) as discussed in Chapter 8.

- Codec usage. In cellular telephony, the use of codec is very closely tied to the cellular network. For example the GSM-AMR codec rates match the cellular network transmission "time slots" exactly. Thus, an existing voice subsystem for a cellular phone may not be able to easily accommodate "standard" VoIP audio codecs such as G729ab and G723. While RTP profiles for the cellular codecs (GSM, EVRC) are defined for VoIP, not all VoIP devices will implement them due to their complexity and processing requirements. Thus, when in VoIP mode, a dual-mode handset may fail to negotiate a codec for a VoIP call, or may require a transcoder somewhere in the network.

10.3.2 Convergence Strategies

There are two basic strategies for Wi-Fi voice/cellular convergence with several variations being proposed or prototyped. The first basic strategy, an example of which is being proposed for GSM networks, is an approach where the lower-layer cellular protocols are replaced with IP. The higher-layer cellular network protocols are then tunneled over the IP network. A gateway function at the border between the IP network and cellular backbone is provided to terminate the tunnel. In the case of GSM, this approach is known as Unlicensed Mobile Access (UMA).

The second strategy, being proposed first for CDMA networks but also applicable to GSM networks, is using IMS as a common signaling protocol for both the pure cellular network signaling and the voice-over-Wi-Fi network.

10.3.2.1 UMA

Unlicensed mobile access, as defined in *UMA Architecture (Stage 2) R1.0.43 (2005-425-298),* is

> *"an extension of GSM/GPRS mobile services into the customer's premises that is achieved by tunneling certain GSM/GPRS protocols between the customer's premises and the Core Network over a broadband IP network, and relaying them through an unlicensed radio link inside the customer's premises."*

Under UMA, a dual-mode GSM/Wi-Fi handset uses the GSM network when it is available and no Wi-Fi network is present. When a suitable Wi-Fi network comes into range, however, the phone will switch over to using voice over Wi-Fi. As defined, UMA is not Wi-Fi-specific and is designed, in theory, to use any unlicensed access technology. Wi-Fi and Bluetooth are called out as initial candidates in the UMA Stage 2 specification.

In the case of a UMA/Wi-Fi-capable handset, GSM call signaling and voice compression will be used in both the GSM cell network and the Wi-Fi network. When using the GSM spectrum (referred to as GERAN/UTRAN mode), the phone operates as a pure GSM phone with some additional functionality present to enable GSM to do Wi-Fi roaming. We will discuss this additional component a little later. Once a switch to a Wi-Fi network has taken place, the phone will use Wi-Fi to access the Internet and will set up a secure tunnel back to the GSM core network (this is referred to as UMAN mode in the specification). Over this tunnel, pure GSM signaling messages and RTP encapsulated media packets will be sent and received. This is illustrated in Figure 10.1.

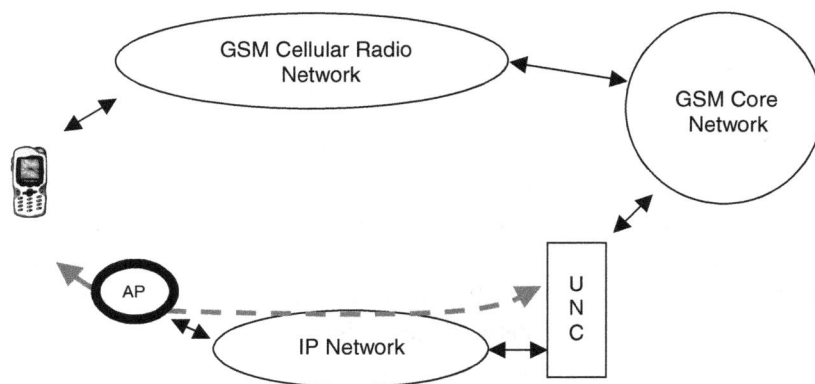

Figure 10.1: UMA Overview

This tunnel is terminated at a secure gateway (SGW) component of a special network device known as the UMA network controller, or UNC. The UNC/SGW sits at the border between the GSM core network and the Internet, and acts as a gateway between the VoIP and GSM world. On the VoIP side, the UNC looks like the endpoint of a TCP/IP connection. On the GSM side, the UNC looks like a GSM base station. The UNC routes data (call-signaling messages or media packets) between the IP network and GSM networks. There are two kinds of UNCs: provisioning UNCs and serving UNCs. The provisioning UNC is used for initial phone bring-up, and will typically redirect the phone to a serving UNC. The FDQN of the provisioning UNC and its associated secure gateway will be typically provisioned into the dual-mode phone.

The tunnel between the dual-mode phone and the UNC will be secured using the standard internet security protocol, IPsec (which we discussed briefly in Chapter 7). The specification calls for the use of the IPsec Encapsulating Security Protocol (ESP) in tunnel mode with AES encryption (Cipher Block Chaining Mode), and SHA1 authentication. Figure 10.2 illustrates the protocol layering of UMA for voice and signaling, respectively.

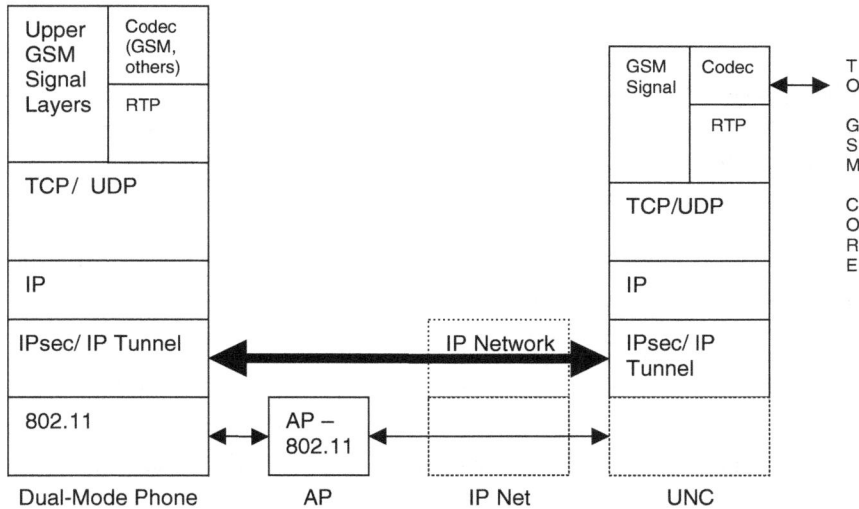

Figure 10.2: UMA Signaling Secure Tunneling

The IPsec security association is set up via the IKE key-management protocol(v2). UMA defines two IKE profiles, one using EAP-SIM and one using EAP-AKA (authentication and key agreement). Both these profiles allow for fast reauthentication. This is useful to reduce the workload due to full IKE v2 handshaking and to speed up the registration process, especially if the dual-mode, UMA phone has roamed onto a new Wi-Fi network (where its assigned IP address has been changed).

It is important to note that this secure tunnel is used for all data between the dual-mode phone and the UNC, including voice traffic. We will have more to say about the bandwidth efficiency of this scheme a little later.

The first operation after setting up the tunnel is to register. This may take several steps and additional tunnels, especially if this is the first time that the phone has booted because of the serving UNC discovery procedure. The procedure consists of the following steps:

- Discovery of the default serving UNC. This will be provided by the provisioning UNC.

- Registration with the default serving UNC

- The default serving UNC may accept the registration, or may redirect the phone to use a different serving UNC.

- In the latter case, the registration will be repeated to the new serving UNC. This will involve setting up another secure tunnel.

- The registration is completed by the dual-mode phone sending a Register Accept message.

The FDQN of the serving UNC/SGW can be saved for subsequent reboots along with the AP BSSID. On subsequent reboots, the phone can attempt to register directly with the serving UNC that it had previously used when connected to this AP. Note that the phone can be redirected to a different serving UNC so that the discover procedure can take place at any time. Also the discovery procedure may be necessary if the phone roams to a new Wi-Fi network where it has not previously operated.

Once the secure tunnel is in place, and the dual-mode phone has registered with the UNC, the dual-mode phone can use the tunnel to transport the higher layers of the GSM call-signaling protocols.

The UNC gateway function translates between the GSM core network and the secure tunnel. The tunnel is also used for the media traffic flow. While VoIP allows for a variety of codecs to be used, UMA uses the cellular standard GSM AMR or WB-AMR codecs (RFC 3267). This is preferred since a call might have roamed from or may in future roam to the cellular network. The codec samples are RTP encapsulated before being transmitted through the secure tunnel. One comment on this is that use of IPsec and secure tunneling introduces substantial overhead to each voice packet. Given that a 20-ms GSM voice frame (at the full AMR rate of 12.2 kbps) will contain 244 bits of speech payload plus 12 bits of frame overhead for a total of 32 bytes, we can compute that a UMA-secure/tunneled RTP packet will effectively be 126 bytes plus layer 2 headers (see Table 10.2).

Table 10.2: Effective GSM RTP Packet Size in UMA Tunnel (with AES Encryption in CBC Mode, SHA1 Authentication)

Packet Element	Size (bits)
Frame payload (12.2 kbps rate)	244
CRM	4
Table of contents	6
RTP padding	2
RTP header	96
UDP/IP	224
2nd IP header (tunnel)	160
IPsec ESP header	32
IV	128
Padding	0
Trailer	16
Authentication	96
Total	**1008**
% Overhead	**321%**

A dual-mode, UMA phone can be set up in one of four preferences:

1. GSM only (i.e., never use the UMAN mode of operation).

2. GSM preferred (i.e., use GERAN/UTRAN mode where possible, switching to UMAN mode only when the GSM network is not available).

3. UMAN preferred (i.e., use voice over Wi-Fi where possible).

4. UMAN only (i.e., switch to UMAN mode immediately after the phone starts up and registers on the GSM network).

The procedure to switch from GERAN/UTRAN mode to UMAN mode (GSM to Wi-Fi handover) is referred to as "rove in" and the reverse procedure is referred to as "rove out." Unlike the inter-802.11 roaming that we discussed in Chapter 8, these two roaming procedures are of the "make before break" type, meaning that the new mode must be fully established before the old mode is disconnected. This is important because, as we have seen before, there are multiple protocol steps in the secure tunnel and registration procedures before voice can actually be delivered to/from the handset. UMA has the goal of seamless switchover between the two modes. It will be interesting to see if this achievable in practice. One problem area will be the delay introduced by the RTP jitter buffer when in UTRAN mode. When in GERAN mode, the phone will be operating without any jitter buffer (this is not required due to the TDMA protocol used in the GSM network). As soon as the UTRAN mode is enabled, the initial RTP packets from the core network will need to be delayed in the handset so that the handset jitter-buffer can be primed. Thus, the user will potentially hear a gap in the conversation equal to the jitter buffer nominal setting. One way around this, potentially, is to begin with a shallow jitter buffer and let it adapt aggressively if network conditions require a deeper buffer. The corresponding jitter buffer in the UNC will need the same kind of work-around.

The UMA specifications include recommendations for the Wi-Fi network that a dual-mode, UMA phone will utilize. These recommendations include:

- Use of 802.11 security, WEP or WPA PSK.

- QoS: Use of WMM is recommended. The specification suggests "simulating" WMM if the AP does not support the feature, by using a "non-standard," smaller backoff window and interframe delay. The specification also calls for the dual-mode phone to use link level 802.1D priority markings and/or the IP layer TOS/DCSP value of received packets for outgoing packets.

- Power save: The specification calls for the use of 802.11 power save when not in an active call but, interestingly, states that the 802.11 power save should not be used during a call. Presumably this was written before the voice-friendly U-APSD power-save mode was defined by 802.11/Wi-Fi.

- Roaming and scanning: The specification calls for background scanning at an undefined interval, "depending on power conservation strategies." The specification recommends the use of RSSI as the key metric to determine when to roam. The specification also states that Wi-Fi roaming is to be isolated from the upper layer protocols, unless the IP address has changed as a result of the roam. For inter-BSS roaming, UMA suggests a target of 100 ms for the device to switch to a new AP.

- From a hardware capability point of view, the specification requires the physical characteristics shown in Table 10.3.

Table 10.3: Wi-Fi Physical Characteristics Required for UMA

Characteristic	Specification
Transmit power (at antenna)	+17 dBm
Receive sensitivity	–87 dBm @ 1 Mbps
Antenna gain	–0 dBi

- Finally, the specification recommends the use of "intelligent" packet loss concealment algorithms to mitigate packet loss.

On the AP side, the specification recommends a beacon period of 100 ms.

UMA, when operating in UMAN mode, has provisions for some amount of RTP session negotiation. UMA allows the following call parameters to be "negotiated," via information elements in the tunneled signaling packets:

- RTP port (UDP)

- RTCP port (UDP)

- Sample size (VoIP packetization period)

- Redundancy/mode tables (see below)

- Initial GSM codec mode to use

- RTP dynamic payload type to use for the audio codec

These parameters can also be changed mid-call through tunneled signaling messages.

The GSM codec has a built-in redundancy mode that is applicable to transport over Wi-Fi and over IP in general. UMA has provisions to take advantage of this feature. The feature works as follows:

GSM inherently supports various modes of operation (or bit rates), ranging from 4.72 kbps to 12 kbps for the narrowband AMR codec (additional rates are available in the wideband, WB-AMR codec). Any of these rates can be used in a call, and the RTP packing format for GSM AMR includes a field (Codec Mode Request or CMR) with which a receiver can signal the other side that he desires a codec rate change. Furthermore, an RTP GSM AMR packet may contain redundancy in the form of copies of previously transmitted GSM frames. This is referred to as forward error correction. UMA allows a redundancy/mode table to be exchanged via information elements in the tunneled GSM/UMA call-signaling packets. This table gives, for each rate, the desired redundancy level (UMA restricts the options to none, one level or two-level). A separate table can also be present that defines, for each mode, the frame-loss rate threshold and a hysteresis level to control when a receiver should try to switch rates. Armed with these tables, a UMA handset can monitor the frame loss it is seeing and, when configured loss thresholds are hit, the handset can use the CMR field to request a rate/redundancy change. Similarly, the receiver in the UNC can do the same. As a note, it is unlikely that a rate change alone will accomplish much when in UMAN mode because of the packet protocol and security overhead mentioned above. However, the switch to a lower bit rate with redundancy has the effect of protecting for packet loss without increasing the overall RTP packet size. This is an important consideration for Wi-Fi QoS networks with admission control and also has a slight impact on power consumption when operating in the Wi-Fi network.

The GSM codec also has a feature known as unequal bit error detection and protection. This is accomplished by organizing the codec payload bits into three classes: A, B and C. Class A bits are the most important bits, Class B are next, and Class C are the least important. For example, in the GSM AMR 12.2-kbps rate, 81 bits out of the total 244 bits in a 20-ms frame are deemed Class A. Thus, in theory, a received packet with corruption in the Class B or C area of the payload could still be used (and not completely dropped). This scheme is, of course, very useful in GSM cellular networks where the packet integrity checks are adjusted to reflect the payload bit classes. Unfortunately, when operating in UMAN mode over a Wi-Fi/IP network, this codec feature is not applicable (although it would be beneficial). The problem is that, first, a UDP checksum covers the entire UDP packet so that UDP checksums would need to be completely disabled for the scheme to work. Even more damaging is the use of the IPsec-protected tunnel. IPsec performs a message authentication check across the entire payload. Thus, bit errors in Class B and C areas would result in the packet being dropped due to authentication failures. Finally, and most damaging, the 802.11 link-level security authentication checks (if enabled) would also fail for the same reason. Using this feature would require "application" knowledge to be propagated down to all layers of the protocol stacks; clearly this is not a feasible approach, at least for the near future.

A final area of interest with UMA is its handling of 911 emergency calls. A UMA dual-mode phone can first be configured as part of the registration process as to which network is preferred to make emergency calls. Secondly, the UMA call-setup message includes an information element to indicate the type of call. "Emergency" is an option in this IE. Finally, UMA has several options for managing location information:

- UMA handsets can send the identifier of their attached AP. The UNC can then use this to "look up" the APs location when an emergency call is placed. This, of course, is only 100% accurate if every possible AP (that UMA allows—UMA has provisions to restrict the APs which the dual-mode phones can use to obtain UMA service) has been registered so that its location is in a back-end database.

- UMA handsets can send their own location if they know this from other means—e.g., GPS.

10.3.2.2 IMS

The second approach to dual-mode telephony is the use of VoIP in both the cellular and Wi-Fi modes, via IMS. The IP Multimedia Subsystem is a key element of the third generation (3G) architecture. 3G is, briefly, is a collaboration of various standards bodies to define the next (third) generation cellular networks. 3G is a unification of the cellular world and the Internet, and IMS is the mechanism that enables IP-level services.

IMS is based on SIP. As we have discussed earlier, SIP is now the main VoIP call-signaling protocol. The reasons for changing from conventional cellular signaling to a system based on SIP are beyond the scope of this book. The driving factor is unification of services, with the idea being that IMS-based signaling can facilitate the deployment of voice, video, presence and other services. However, the use of SIP/IMS and VoIP over the existing cellular networks has some interesting technical challenges that, if solved, will play into a pure voice-over-Wi-Fi scenario as well.

One issue is bandwidth. Today's cellular networks are constrained as to the amount of bandwidth available for a voice call. Recall from Chapter 3 that SIP-based VoIP call signaling utilizes a relatively inefficient, text-based protocol to communicate call signaling information. Furthermore, if you look at a VoIP media packet, a good portion of this packet will be composed of packet header information. Thus, IMS signaling and media will require more bandwidth than the current cellular protocols.

In the case of SIP messages, one approach is to use compression techniques such as those defined in RFC 3320. With RFC 3320, a layer that performs lossless compression (e.g., gzip) can be inserted between the application (voice-signaling SIP stack) and the network protocol stack (TCP/IP). On transmission, this layer can run a native implementation of a compression algorithm. However, on reception, this layer makes use of the Universal Decompressor Virtual Machine (UDVM), which is essentially a JAVA-like virtual machine tailored specifically for decompression operations. The instructions or "byte-codes" to be executed are provided by the sender. The advantage of this approach is that the actual compression algorithm can be controlled entirely by the transmit side; it can pick any algorithm desired, perform the compression on a message, and send it along with the UDVM instructions on how to decompress it (the bytes codes would only need to be sent with the first message, assuming the same algorithm is used throughout the signaling session).

To tackle the problem with the media packet protocol overhead introduced by the RTP/UDP/IP headers, an approach is to use a technique called robust header compression (RHOC- RFC 4362). RHOC can work across a link layer (e.g., in 802.11 between the phone and AP). The basic idea behind RHOC and its related header-compression protocols is to define at the start of a packet flow (e.g., at call setup), which fields in the packet headers are static, which fields update according to simple rules (e.g., the RTP timestamp), and which fields need to be sent along with each packet. Once these are set up, the static fields and those that change in a simple way can be stripped before packet transmission. The receiver will then reconstruct the complete header before forwarding the packet. The protocols include mechanisms to recover from delivery problems—for example, if a burst of packets is lost for some reason, the preset header information may need to be changed.

10.4 WiMax

WiMax is a new wireless technology, defined by IEEE 802.16x standards. The core standard, 802.16, defines protocols for a broadband wireless infrastructure, operating in the 10–66 GHz frequency range. The basic topology defined in the specification is point-to-multipoint. The targeted data throughput range was 70 Mbits, with a peak rate of 268 Mbps and a typical cell radius of 1–3 miles. This base standard was subsequently enhanced with a suite of amendments known as 802.16a. These added considerations for more spectrum bands (licensed and un-licensed), support for non-line-of-sight architectures, new physical-layer specifications and enhancements to the MAC layer. These later changes included consideration for quality of service and different types of traffic, including voice.

A second version of WiMax is currently being defined. This version, based on the 802.16e specification, is addressing mobility and roaming considerations. It will include support for hard and soft handoffs and improved power-saving techniques. It introduces a new PHY layer optimized for mobility.

Like 802.11, the 802.16 specifications include multiple physical layers. 802.16a defines three protocols:

- A single-carrier modulation format.
- Orthogonal Frequency Division Multiplexing (OFDM), with a 256-point transform. This is the same modulation technique used in 802.11g.
- Multiuser OFDM (OFDMA), with a 2048-point transform.

802.16 adds a new variant of OFDMA, referred to as SOFDMA (the "S" stands for scalable). The variant provides better performance for multiple users under varying conditions.

The media access layer for 802.16 is quite a bit different than for 802.11. It is based closely on the data-over-cable specification (DOCSIS). The relationship between WiMax and Wi-Fi is still to be defined. The conventional school of thought is that WiMax will become a "last-mile" technology, providing an alternative for the currently deployed broadband technology (i.e., DSL, cable, fiber-to-the-home, etc.). A WiMax CPE device, for example, could contain an 802.11 subsystem as well, just like today's cable and DSL broadband routers. With this architecture, a voice-over-Wi-Fi phone would be another subscriber to the WiMax backbone.

802.16e, with its support for mobility, muddies the waters. Conceivably, 802.16e could be used as a replacement for Wi-Fi in some environments.

10.5 VoWi-Fi and Bluetooth

Bluetooth (BT) is radio technology geared at the 2.4–2.5835-GHz ISM unlicensed frequency band just like Wi-Fi (802.11 b/g). Table 10.4 summarizes 802.11/Wi-Fi and Bluetooth technology.

Table 10.4: Wi-Fi (b/g) / Bluetooth Comparison

Wi-Fi	Bluetooth
Direct Sequence Spread Spectrum (DSSS).	Frequency Hop Spread Spectrum (FHSS).
Use only 22 MHz × 3 (channel) = 66 MHz.	Use 1 MHz × 7 9 (channel) = 79 MHz. The hop rate is 1600 hops/sec.
Power: 1 W (30 dBm).	Power: 1 to 100 mW.
Data rate: up to 56 Mbps at close ranges, 5.9 Mbps < 175 ft, 5.5 Mbps at 250 ft.	Max data rate 550 kbps at 250 ft.
Range up to 100 meters, depends on power and environment.	Power range: 100 meters (class 1), 10 meters (class 2), 10 cm (class 3).
Each Wi-Fi network uses 1 channel, max 3 nonoverlapping networks.	FCC requires BT devices to hop ≥ 75 channels up to max 79 channels.
Defined only layer-2 protocols, a common way to access Internet through the AP.	Define different layers of protocols. Profiles allow for different voice and data application interworking.
Security is in the layer 2 (WEP, WAP, WAP2, etc.).	Security is in layer 2 (LMP) and security architecture for different layers.
Allow one AP and many stations to bind.	Allow one pair of AG (Audio Gateway) and handset or hands-free to pair.

There are three classes of BT devices, each with a different maximum output/power and corresponding range profile. These are summarized in Table 10.5.

Table 10.5: BT Device Power Classes

Power Class	Max Output Power (mW)	Maximum Range (meters)
Class 1	100 mW	100 m
Class 2	2.5 mW	10 m
Class 3	1 mW	10 cm

The current uses of BT (at least in class 2 and 3) make it a complementary technology to Wi-Fi. In the context of a VoWi-Fi phone, a BT subsystem might be present to provide low-rate, close-proximity wireless access to peripherals. The most likely scenario is where a VoWi-Fi phone would have a BT subsystem to allow the use of a BT handset or handsfree device as the end audio transducer.

This leads us to the main issue with BT and Wi-Fi: coexistence. In a nutshell, the BT physical layer utilizes a frequency-hopping technique that unfortunately can cause interference in an 802.11 b/g network (and vice versa). As described earlier, Wi-Fi/802.11 b/g standards in North America divide the ISM band into 11 overlapping channels (In Europe and Japan additional channels may be present). Only three channels—1, 6, 11—are nonoverlapping. Each channel utilizes 22 MHz of the ISM band; thus $3 \times 22 = 66$ MHz out of the 88.35-MHz ISM band will be occupied by a fully loaded Wi-Fi deployment.

Bluetooth, on the other hand, uses a frequency-hopping technique across almost the entire ISM band. Each hop frequency is 1 MHz and up to 79 channels are allowed. Furthermore, BT specifies a hop rate of 1600 hops/sec. This means that transmission from a BT device will definitely overlap with Wi-Fi transmissions if it is in range and the Wi-Fi transmission is long enough. This is shown graphically in Figure 10.3.

The impact of the interference from BT devices on Wi-Fi equipment is to effectively raise the Wi-Fi channel bit-error rate. A BT device that wants to transmit will be unaware of Wi-Fi activity and will not delay its transmission. If it is in range and its frequency-hopping scheme happens to overlap the Wi-Fi transmission, the Wi-Fi receiver will see a degraded signal and can either miss the packet or detect a CRC error. In either case, the Wi-Fi transmitter will need to resend.

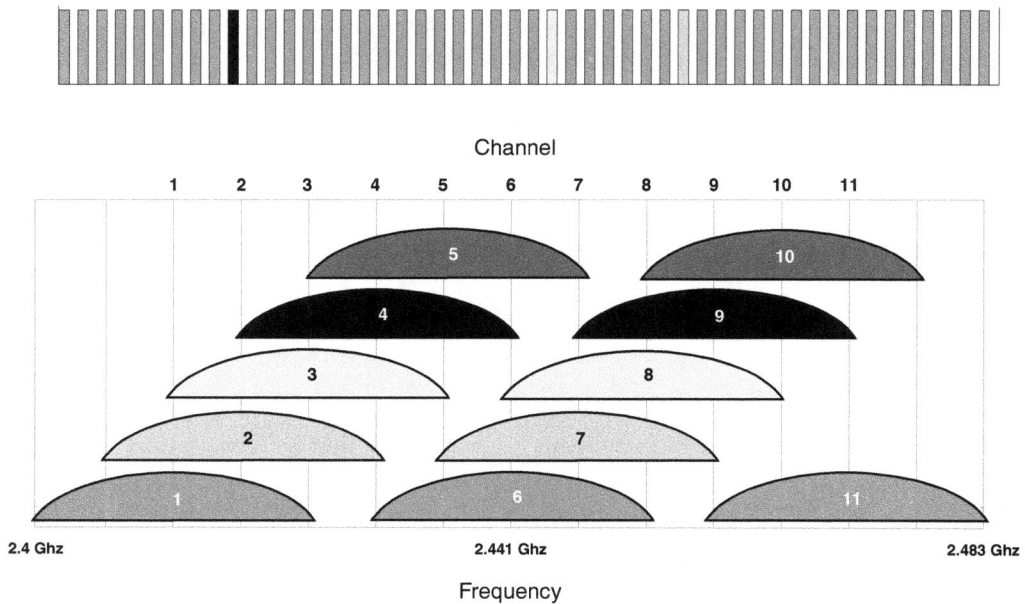

Figure 10.3: 802.11 b/g Frequency Bands

Packet retransmission in Wi-Fi typically will lead to a reduction of transmission rate, as the Wi-Fi devices attempt to react to what they perceive as a noisy environment. Thus, data that would be normally transmitted at 54 Mbps may eventually be transmitted at a rate of 11 Mbps. This reduces the overall throughput of the Wi-Fi network and has other side effects for voice such as increased latency and power consumption.

Figure 10.4 conceptually illustrates the effect of a BT transmitter on Wi-Fi throughput. The figure plots Wi-Fi throughput versus received signal strength for the cases where the BT device is transmitting or not. Received signal strength here is used as a generalization of distance between the Wi-Fi device and its access point. The net effect of a BT transmitter in close proximity is to sharply degrade throughput, even when the Wi-Fi device is close to the AP. The degree of impact of the BT transmitter is correlated to the BT device location to the Wi-Fi device.

As an unpleasant side effect, Wi-Fi packets sent at the lower Wi-Fi rates will stay on the air longer and are hence even more likely to experience BT interference. Note that, because of the frequency-hopping technique used in BT, a Wi-Fi transmitter may not be able to detect that a BT transmission is in progress and back off. This in contrast to the case of Wi-Fi channel overlap; in this situation a Wi-Fi device will be more likely to detect the 802.11 energy and can then back off.

Wi-Fi Throughput

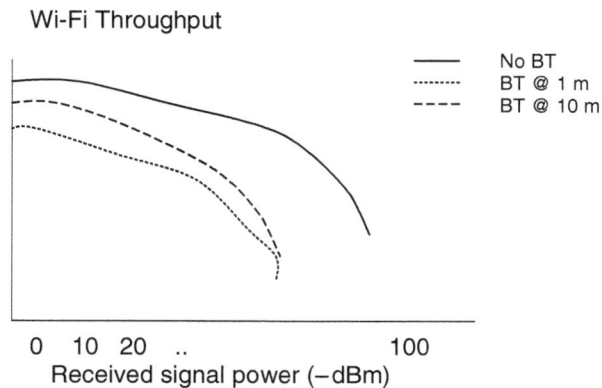

Figure 10.4: Wi-Fi Throughput vs. Received Signal Strength (AP-STA distance)

Before discussing ways for BT and Wi-Fi to coexist, we need to discuss the types of BT connections that can be used. BT has two types of link-level protocols: asynchronous connectionless (also known as ACL), and synchronous connection-oriented (SCO). ACL BT protocols are typically low rate and allow for packet retransmission. SCO connections are higher speed and (in BT 1.0 devices) do not allow for packet retransmission. SCO connections are used, for example, to communicate to BT headsets and hands-free devices. Different coexistence techniques are required for each of these, depending on the type of Wi-Fi traffic. For a VoWi-Fi phone with an adjunct BT handset or hands-free device, we are interested in Wi-Fi voice coexistence with BT voice over an SCO connection.

Several coexistence schemes for BT and Wi-Fi are possible. We will discuss some of these below. It is important to recognize that a combination of schemes will be required for a full robust solution.

One technique, utilized by Wi-Fi and BT chipset providers such as Texas Instruments, is to provide silicon-level interfaces so that the two chipsets can collaborate to minimize interference. In the Texas Instruments solution, for example, its two chip sets share a coexistence interface over which information on when transmission is taking place can be exchanged. If a BT transmission is going on, the Wi-Fi transmission can be delayed and vice versa. This approach works best for Wi-Fi data and BT data (i.e., ACL) coexistence, with a couple of limitations. It is not enough to solve Wi-Fi voice and BT voice (SCO) coexistence problems, however.

A second set of techniques comes from the 1.2 version of the BT standard. This update has taken steps to address the coexistence issue by incorporating two new features: adaptive frequency hopping, and an enhance SCO link protocol (ESCO).

The BT 1.2 adaptive frequency-hopping scheme allows a BT device that has knowledge of the 802.11 device that it is colocated with to adjust its frequency-hopping scheme accordingly.

For example, if a BT device knows that its 802.11 counterpart is operating on Channel 1, it can select frequencies out of the 802.11 channel 1 subband for its hopping sequence. Thus, the BT device would use 57 out of the possible 79 channels. Using this technique in practice for the case of VoWi-Fi and a BT headset/hands-free device has several issues to overcome:

- The BT 1.2 specification does not define how the BT device learns the channel use. Typically this would require a software interface between the Wi-Fi and BT chipset/ device driver.

- The adaptive frequency-hopping scheme does not help as much in cases where all multiple 802.11 channels are in use, such as would be the case in an enterprise environment. The 1.2 compatible BT devices can skip around the 802.11 channel that the colocated Wi-Fi is actively using, but still may interfere with other Wi-Fi devices as the user roams.

- Furthermore, in a multiple AP environment the BT device will need to change its hopping sequence whenever the Wi-Fi device decides to roam. This will most likely result in an interruption in the BT data stream.

- There will be impact on the scanning techniques discussed in Chapter 8. For example, the use of unicast probes to discover new APs on other channels will be problematic.

- A final issue is that a combined Wi-Fi/Bluetooth device will have cost pressures to share a single antenna. The above techniques are appropriate if each subsystem has a dedicated antenna and there is a minimal degree of RF separation between the two. When the antenna is shared, it is unlikely that the frequency-hopping adjustment approach will be effective.

ESCO allows for higher speed and retransmission on the SCO links. This will improve the quality of the BT transmissions (e.g., voice to/from a BT handsfree device).

In short, coexistence for BT voice and VoWi-Fi is still an open technical challenge.

10.6 VoWi-Fi and DECT

We have left this topic near the end as it is perhaps the most controversial, especially to DECT proponents. Digital Enhanced Cordless Telecommunications is a popular wireless standard, mostly in Europe, that—it can be argued—provides a complete wireless telephony solution today. DECT works in the 1.9-GHz band (some versions are available in the 2.4-GHz band) and utilizes a time-division multiplexing approach to bandwidth allocation. It was primarily geared for cordless telephony. An overview of DECT is given in Table 10.6.

Table 10.6: DECT Summary

Characteristic	DECT
Frequency Band	1.9 Ghz
Access Method	TDMA
Data Rate	2 Mbps – being expanded to 20 Mbps for data services
Range	50 meters indoors, 300 meters outdoors
Modulation	Gaussian Minimum Shift Keying (GMSK)
Voice Codecs	G726 (ADPCM) [32 kbps]
Voice Signaling	ISDN based
Handovers/Roaming	Built into protocol
Security	GSM based
Cost	Approx ½ compatible 802.11 solution
Battery Life	~ 12 hrs talk, 100+ standby

In this regard, DECT can be considered a competing technology to VoWi-Fi. Let's identify the advantages touted by DECT adherents (many of these are based on original, 802.11b voice-over-Wi-Fi implementations)

- Cost: DECT handsets are cheap! This is based partly on the maturity of the technology, so that highly integrated hardware solutions are available.

- Power: DECT was designed upfront to be power efficient. It utilizes a TDMA-based method. DECT receivers can shut off their radios until their time slot occurs.

- Handset performance: Again, because of the maturity of the DECT handset market, industrial-strength (shock, temperature, dust, etc.) equipment is available today.

- Handoffs: DECT was designed with handoffs in mind.

- Range: DECT has inherently better range than 802.11. 802.11 can extend range by adding repeaters or access points, but this adds to cost and has limitations based on the ISM channel bandwidth.

- Quality of Service: The original DECT objections to Wi-Fi QoS (or lack thereof) were based on 802.11b deployments.

- Security: Most DECT objections to Wi-Fi security are based on the WEP implementations. As we have seen, WPA and WPA2 have addressed these concerns. However, as we also have seen, the use of WPA and WPA2 authentication and key-distribution methods makes fast handovers more complex.

The bottom line is that, while DECT does have advantages over VoWLAN for pure telephony, these are due primarily to its inherent limitation of being primarily a telephony protocol. As a "telephony-first" protocol, DECT will naturally win in a phone-only environment. But if we add data to the mix, voice over Wi-Fi will be a more attractive solution. Wi-Fi is the clear winner for providing wireless data service. Voice over Wi-Fi, as it runs on top of the data network, will succeed just as pure voice over IP is.

The other issue with DECT is how it plays into a VoIP backbone. DECT uses a ADPCM codec between the handset and the base station. This requires ADPCM transcoding if another low bit-rate codec is to be used for the network portion call. Furthermore, DECT uses ISDN-based signaling between the base station and the phone. This will need to be translated into SIP VoIP signaling.

We can also look at DECT and Wi-Fi in another light, that of convergence. There are various projects underway to merge DECT and Wi-Fi together. One approach has been to integrate the upper layers of DECT with the 802.11 MAC and PHY.

10.7 VoWi-Fi and Other Ongoing 802.x Wireless Projects

In this section we will take a quick look at three ongoing IEEE wireless standards and their potential relationship with VoWi-Fi.

10.7.1 802.20

The mission of the 802.20 project (also referred to as Mobile Broadband Wireless Access or MBWA) is to "develop the specification for an efficient packet based air interface that is optimized for the transport of IP based service." There is a special emphasis on mobility in this project, with goals of handling subscribers moving at speeds of up to 155 miles per hour (e.g., for high-speed train service). By this definition, there is overlap somewhat with the goals of 802.16e. However, the scope of 802.20 is limited to below the 3.5-GHz band, while 802.16e covers additional spectrum. Also, 802.20 is targeting a much lower data rate (around 1 Mbps) than 802.16.

As an IP-based service, 802.20 must deal with similar issues as 802.11 when used to carry VoIP.

There is a lot of debate on how 802.20 and WiMax (802.16e) will evolve, since there is a great deal of overlap. It is possible that 802.20 will be restricted to the high-speed domain only.

10.7.2 802.21

The 802.21 is an interesting project with special relevance to the voice application. Its goal is to "develop standards to enable handover and interoperability between heterogeneous network types including both 802 and non 802 networks." In other words, the project is involved with standardizing the type C and D roaming that we discussed in Chapter 8. This is also referred to as Media Independent Handoff, or MIH. Among the topics that 802.21 is investigating is the definition of a common interface between various layer 2 (802.11, 802.16, etc.) and layer 3 to facilitate the roaming and handoff process.

The draft standard discusses the concept of link-level "triggers." These are link-level events that can be passed to layer 3 to provide link state information to the roaming decision process. Link-level triggers include such events as link up/down, link quality above or below a defined threshold, link QoS state, perceived link range or throughput, and even network cost. In a pure 802.11 network, the link-level triggers correspond to the VoWi-Fi device's roaming triggers that we discussed in Chapter 8, such as the RSSI, beacon miss rate, and retransmit rate. However, in 802.11, these triggers were not explicitly called out as such and their use is up to the device manufacturer. The 802.21 project attempts to define these and to provide a framework for their configuration and reporting.

One difference in the 802.21 framework is that it includes the idea that triggers can come from the remote side of the connection, as opposed to being generated solely by the local side. In an 802.21-enabled 802.11 network, for example, the AP would be able to use a layer-2 message to send a "suggestion" that the station roam.

Another aspect of 802.21 is support for the "make before break" concept. As we discussed in Chapter 8, 802.11 today requires that a station disconnect from one AP before connecting to another ("break" before 'make"). This approach has some drawbacks, especially for voice, because there will be a period of outage between the "break" and the subsequent "make." As we saw in Chapter 8, we can mitigate some of this latency in a pure BSS roaming situation through such techniques as WPA2 preauthentication. However, with dissimilar network roaming, the latency in security setup, IP address provisioning, etc. will be too long for the goal of seamless mobility.

802.21 also introduces the idea of a mobility-management service. This is a network-based service that mobile devices can register with to obtain information about other networks that could be roaming candidates. This is somewhat analogous to the proposed 802.11k AP list that we discussed in Chapter 8, but it covers not just APs, but also cellular base stations, WiMax head ends, and so forth.

10.7.3 802.22

The 802.22 project is working on how to use portions of the RF spectrum, currently allocated to television broadcasting, for carrying wireless data services. In particular, the UHF/VHF TV bands between 54 and 862 MHz are being targeted, both specific TV channels as well as guard bands (white space). This standard is still under development, but it looks to be based on 802.11 protocols, possibly adding an additional PHY layer and enhancing the MAC layer to deal with longer range.

One interesting aspect of proposed 802.22 networks, also referred to as wireless regional area networks (or WRANs), is that they will utilize a new technology known as *cognitive radio*. The proposed ITU definition of this technology is: "a radio or system that senses and is aware of its operational environment and can dynamically, autonomously, and intelligently adapt its radio operating parameters." The basic idea is to allow the wireless nodes to manage the spectrum in a distributed fashion, by observing the environment.

10.8 Conclusion

This chapter has taken a look at the future of voice over Wi-Fi, and how it may evolve, coexist and interact with other wireless technologies.

References

[1] Camarillo, G. and M. Garcia-Martin, *The 3G IP Multimedia Subsystem*, John Wiley, 2004.

[2] *Fixed, nomadic, portable and mobile applications for 802.16-2004 and 802.16e WiMAX networks*, November 2005, prepared by Senza Fili Consulting on behalf of the WiMAX Forum.

[3] IEEE Standard 802.16: *A Technical Overview of the Wireless MAN™ Air Interface for Broadband Wireless Access*

[4] *Unlicensed Mobile Access (UMA) Protocols (Stage 3)*, R 1.0.4, 5/2/2005.

[5] "Global, Interoperable Broadband Wireless Networks: Extending WiMAX Technology to Mobility," *Intel Technology Journal*, August 20, 2004.

[6] "Scalable OFDMA Physical Layer in IEEE 802.16 WirelessMAN," *Intel Technology Journal*, August 20, 2004.

[7] A Generalized model for Link Level Triggers, V Gupta, et al., [802.21 Contribution]

[8] *http://www.comsoc.org/oeb/Past_Presentations/CityWiFiMesh_Apr04.pdf*

[9] RFC 3267 Real-Time Transport Protocol (RTP) Payload Format and File Storage Format for the Adaptive Multi-Rate (AMR) and Adaptive Multi-Rate Wideband (AMR-WB) Audio Codecs, J. Sjoberg et al., June 2002.

[10] RFC 4362 RObust Header Compression (ROHC): A Link-Layer Assisted Profile for IP/UDP/RTP, L.E. Jonsson et al., December 2005.

[11] RFC 3320 - Signaling Compression (sigcomp), R. Price et al., January 2003.

Index